THE PS ROYAL WILLIAM OF QUEBEC

THE PS ROYAL WILLIAM OF QUEBEC

The First True Transatlantic Steamer

EILEEN REID MARCIL

ISBN 978-1-77186-229-5 pbk; 978-1-77186-251-6 epub; 978-1-77186-252-3 pdf

Book Design and cover by Folio Infographie
Editing and proofreading by Robin Philpot, Blossom Thom

Front cover illustration: *PS Royal William*, Musée national des beaux arts
du Québec (L74.260). Photograph: Patrick Altman
Back cover illustration: The *Royal William* Leaving Pictou. Print. Adam Sherrif Scott. M16738 McCord Museum

Legal Deposit 4th quarter 2020
Bibliothèque et Archives nationales du Québec
Library and Archives Canada

Published by Baraka Books of Montreal

Printed and bound in Quebec

Trade Distribution & Returns
Canada – UTP Distribution: UTPdistribution.com

United States
Independent Publishers Group: IPGbook.com

We acknowledge the support from the Société de développement des entreprises culturelles (SODEC) and the Government of Quebec tax credit for book publishing administered by SODEC.

Société de développement des entreprises culturelles Québec

Funded by the Government of Canada
Financé par le gouvernement du Canada
Canada

CONTENTS

Acknowledgments

When our son Tony was five years old, his granddad gave him a model steam engine and showed him how it worked. I have often wondered whether he did not intentionally sow a seed in his grandson that grew into a degree in mechanical engineering. I must admit that any thoughts of steam and its applications that I may have harboured at the time, did not linger for long. Some forty years later, however, as a maritime historian in Quebec, I was inevitably confronted with the paddle steamer *Royal William* controversy and knew that the day would come when I would attempt to put it to rest.

I began what became many years of research by consulting Brian Hallett, who was then an archivist at the Public Records Archives in Ottawa. He opened a drawer, took out a file, and handed me a sheet of paper, on which there were several references to the *Royal William* he had collected over the years. He had felt sure, he said, that one day someone would want them. I thanked him for the running start they gave me and remain grateful to him now.

One entry on his list, 'Model - Runge Press,' particularly intrigued me, and thanks to Brian, and the generosity of Fred Runge, whom I traced then and again later, and with ground work from Karen Molson, that model of the PS *Royal William* can now be seen in its permanent home at the Literary and Historical Society of Quebec.

Genealogical research then led me to the home near Chicago of Ainslie Goudie, grandson of James Goudie, who superintended the construction of the *Royal William;* and from there to his descendants, Peggy Bolenbaugh and Grace Budway in California, from whom I would learn more.

I met Campbell family descendants, also: Nora Hague of Montreal, David Wurtele Eaves of Laval, Quebec, and Peter Engler, of Perth, Western Australia, who all have a familial connection with the Campbell & Black shipyard in Quebec where the *Royal William* was built, and were generous in sharing their knowledge.

Other information came from Fred L. Colby Jr and Donald Holmes, regarding their ancestors, John Lowe and Louis Shickluna respectively, both of whom participated in the building of the *Royal William*, and from Dr. John Berry of Newport-on-Tay, Scotland, who allowed me to delve into the letters in his family archive there. Later, Elizabeth Wickstead, of London, UK, would send photographs of the two Wicksteed brothers.

One by one, the descendants and other members of the families of men closely identified with the *Royal William,* put life into the steamer's story for me, and I hope that they will do the same for you.

Another path took me to the museums and archives that hold the specifications, contracts or plans from which the *Royal William* was built. Strangely, it began at a museum in British Columbia, on the other side of Canada, which holds the specifications, 'as referred to in the contract,' for the steamer's hull, signed by William Stevenson, then the acting secretary of the owner company. Others I found at the Bibliothèque et Archives nationales du Quebec, and at its branch in Montreal. Plans for the hull of the steamship were brought from Scotland by James Goudie who revised them in Quebec. Included is a copy of the three-foot-long original plan he gave to the Chicago Historical Society in 1870, a donation widely reported on both sides of the Atlantic. While a copy of the photos of the drawings of the engine, came from Fred Colby, who deposited the originals at the Dossin Great Lakes Museum in Detroit.

The history of the *Royal William* has a strong Nova Scotian connection, and I benefitted from the invaluable technical assistance and interest in the project of David A. Walker, naval architect, marine surveyor, and historian, of Halifax, who gave of his greatly appreciated time to help. In addition, research I needed at the Nova Scotia Legislative Assembly and other archives in the province was kindly undertaken by Marven Moore, curator emeritus of the Maritime Museum of the Atlantic. I am grateful to them both and thank them also for agreeing to read my manuscript. I also thank Philip Hartling and John MacLeod, of the Nova Scotia Archives, with regard to the *Royal William*'s Pictou registration. Finally, I was delighted when Winfried Viebahn, Site Manager of the Uniacke Estate Museum Park, in Mount Uniacke, Nova Scotia, responded to my query with the portrait of Richard John Uniacke on page 26.

I am indebted, also, to Barbara Campbell, of Bexhill, Sussex, who carried out research at the Naval Museum in Madrid for me long ago, opening another avenue of research. I had hoped to follow it with a visit to the museums and archives in Spain, but life dictated otherwise. So, I have relied on the account in the 'Historic Epilogue' of Luis Delgado Bañón, Spanish naval captain and historian's book, *El vapor de ruedas Isabel II*, for the true ending of the remains of the *Royal William/Isabel II.* My thanks are also due to Francisco Salguero Hernàndez of the Librería-Editorial Náutica Cartamar, La Coruña, Spain, for his interest and help.

Nor would the acquisition of a copy of J.W. van Nouhuy's *De Eerste Nederlandsche Transatlantische Stoomvart in 1827 Van Zr Ms Stoompakket "Curaçao"* have been useful had it not been for the translation of excerpts that was kindly undertaken by Willeke Blanchet of Quebec and Willem Bakker of Winnipeg, for which I thank them.

From the Toronto Reference Library's rich collection of prints, and thanks to the gradual lifting of the COVID-19 restrictions and the efforts of the team responsible for carrying it out, in particular Christopher Coutlee and Beau Levitt, I am happy to be able to include the library's fine engravings in chapters 1 and 5 and the Epilogue. How ironic it is to be writing in the midst of a pandemic in 2020 about the *Royal William*'s crossing, which was itself the result of a cholera pandemic reaching Quebec in 1832.

As for the latter part of the *Royal William*'s story, included in the listing of the Kivas Tully papers at the Toronto Public Library and dated 1875-1895 is a scrapbook of "Particulars in relation to the Steamship 'Royal William' which sailed from Quebec August 5th 1833." It is brimming with information that the archivists Anna Zanardo and Christina Szollosy, and more recently Anna Skrynnyk, were only too willing to disseminate, and when I considered a follow-up visit to Toronto, Tony, who lived there, suggested that he investigate first. He photographed the documents that he thought would interest me in the collection, and later we visited the Bibliothèque et archives nationales du Québec to photograph those I needed there. My special thanks to Tony for his interest and for other research he has done for me over the years, as well as for helping me presently with the revision of the text.

Largely because of my membership in the Canadian Nautical Research and other historical societies, I have had the good fortune of meeting many maritime historians on both sides of the Atlantic. In the UK, what more could I have asked for than the friendship and encouragement of two giants of maritime history, Robin Craig of the International Maritime Economic History Association, and Basil Greenhill of the National Maritime Museum at Greenwich, both of whom were interested in Quebec shipbuilding. I am so grateful, too, that the ready advice of Anne Cowne of the Lloyd's Register Foundation carries on.

As well as those in Canada already mentioned, I thank Alec Douglas, of the Canadian Directorate of History, Maurice Smith, of the Marine Museum of the Great Lakes at Kingston, Donald Tremblay, of the Rimouski Naval College, Heather McNabb, of the McCord Museum in Montreal, and Garth Wilson and Sharon Babaian, of the Canada Science and Technology Museum. For their help at Library and Archives Canada, thanks are due to Patricia Kennedy, and more recently, to Marie-Julie Hyppolite and Lynn LaFontaine, while for theirs at the Bibliothèque et archives nationales du Quebec, I am grateful to Annie Labrècque, Pierre-Louis Lapointe, Rénald Lessard, Michel Simard, and in all cases to others on their staff as well.

I thank also Jean-Pierre Charest, Jean Cloutier, Walter Lewis, Ken McKenzie, Larry McNally, historians, Howard

Ross, specialist in computer text formatting and Larry Carey, videographer, from each of whom I have learned.

This book might not have been published had it not been for the prodding of John Fry, editor, historian, and friend, of Katonah, N.Y. and former Canadian, to whom I remain extremely grateful. And, I thank also Helen Meredith for her support.

To Robin Philpot, Publisher of Baraka Books, whose task would turn out to be doubly difficult due to the demands and limitations of the pandemic, a big thank you, Robin, for keeping things going and for the finished result. Many thanks also to graphic artist Josée Lalancette who designed the book, to proof reader Blossom Thom, and to Caroline Soucy, whose skill as an artist and designer of computer images produced the image of the Parliamentary plaque.

My thanks to my children, all of whom, as well as Tony, have helped: Ian who made the map of the *Royal William*'s course and found answers to difficult research problems, Paul, who has read the manuscript and made incisive suggestions more than once, Susie who, despite disruptions to her own schedule has done a great job of looking after my personal needs, allowing me to devote myself to writing, etc. While son-in-law, Daniel Johnson, has also taken an interest in the project and anticipated and dealt with a variety of problems. Among them, getting the attention of a publisher and helping Susie set me up in their home, complete with a computer-equipped study of my own, during the months that COVID-19 was taking its toll in the seniors' residence where I live.

As for my fellow residents, I have benefitted from and thank for their support, Robin Benity, Natalie Fochs, Heather Hodgson, Harvey Hyarosky, Rachel and Edward Keyserlingk, Mary Rankin, Jacqueline Ross, Francisco Sorolla, Magda Turcan, Dorothy and Ronnie Wyer, not forgetting the management and members of the staff, who have continued to care for us with their habitual solicitude.

Although it has not always been on the front burner, it is more than 35 years since I first put the pot on the stove. To others that I may have forgotten, I offer my sincere apologies.

Moreover, due to the lapse since I began the research, several of those that I have mentioned have passed away. I dedicate this book to all who have had a part in it, and cared.

Eileen Reid Marcil
Westmount, Quebec.
2020

Prologue

The world's commerce, especially between North America and Europe, was revolutionized in the 19th century when the propulsion of ships changed from sail to steam. Winds and ocean tides lost their absolute power at sea, while the genius of engineers produced engines that not only replaced them but provided safer and swifter travel, but it took time.

In 1712, an English ironmonger named Thomas Newcomen built the first practical fuel-burning machine that produced steam power. It took the place of the power produced by draft horses that was generally used to pump the water out of flooded mines and to raise the water that powered water wheels. Newcomen's pumps were a boon to mine owners, as the cost of the power produced by a steam-driven pump was far less than its overall cost when it was produced by horses.

Newcomen died in 1729, and during the first Industrial Revolution, from the mid-18th to the mid-19th centuries, Newcomen engines played a major part in the development of the steam-powered factories that gradually replaced many of the cottage industries.

In 1764, a Scot named James Watt, when repairing a Newcomen steam engine, realized that a great deal of steam was being wasted, and spent the next five years improving the issue. He took out a patent for his improvement in 1769, and during his lifetime took out another five. Those and his many other contributions to the practical use of steam power brought Watt respect and fame, which is why, at the International Conference on Electrical Units and Standards, in London in October 1908, the international definition adopted for an Electric Unit was named a "watt."

Between 1801 and 1808, Richard Trevithick, a mechanical and mining engineer, and inventor, from Cornwall, England, known as the harnesser of high-pressure steam, built two steam-driven carriages followed by three steam-driven railway locomotives. On their trials, successes alternated with misfortunes and, frustrated when the locomotives proved too heavy for the cast iron rails on which they were driven, he did not follow through with his intention of building a public railway. Instead, he immersed himself in other projects, including solving problems in the silver mines of Peru.

His railway project was realized in 1825, when the father and son team of George and Robert Stephenson built a steam locomotive that hauled both freight and passengers. Their Stockton and Darlington Railway, which ran through County Durham in England, was the first public steam railway in the world.

Meanwhile, experiments in steam navigation had taken place in many countries. In 1783, Claude François Dorothée, marquis de Jouffroy d'Abbans, a French mechanical engineer and inventor, successfully travelled several kilometres upstream on the River Saône, near Lyon, France, in his 46-meter-long 182 ton steam paddle boat *Pyroscaphe,* acclaimed by thousands of spectators on the river banks. Though he continued to steam on the river for sixteen months, due to a number of untoward circumstances, he did not patent his invention.

Robert Fulton, an American artist and engineer, and his fellow American patron, Robert R. Livingston, a lawyer, politician, and diplomat, would benefit from Jouffroy d'Abbans' original ideas when they met him in Paris in 1801. The partners built two steam driven paddleboats and ran them on the Seine.

Robert Fulton, inventor, engineer and artist, and his partner, Robert R. Livingston, set up the first commercially successful steamboat service on the Hudson River in 1807.

Livingston had acquired the sole right to set up a steamboat service on the Hudson River. They ordered an engine from the Boulton and Watt works in Birmingham, England and in 1807, they inaugurated the world's first commercially viable steamboat service on the Hudson River between Albany and New York. Over the years, Fulton was idolized, but was honest in acknowledging the part that d'Abbans' ideas had played in their success, "if to any one belongs that glory it is the author of the experiments made at Lyons in 1783," he wrote.

By the first years of the 1830s, transport by steamboat on lakes and rivers was no longer exceptional, but long-distance voyages still had to be made in stages. This was to allow for refueling on the way, as so much fuel was required in relation to the vessel's size.

That is why the *Royal William* left from Pictou, Nova Scotia to cross the Atlantic on August 18, 1833, with a load of 254 caldrons of Cape Breton coal, known for its rich bituminous quality. Captain John MacDougall wrote to a friend from London two months after their arrival there, "We were very deeply laden with coal, deeper in fact that I would ever attempt crossing the Atlantic with her again."

Nevertheless, the *Royal William* arrived safely at Portsmouth, England on September 9, 1833, and became the first vessel to span the Atlantic under steam.

CHAPTER I

The Forerunners

IN THE SUMMER OF 1807, something that was happening on New York's Hudson River had John Molson, the Montreal brewer, taking note. His interest would grow when he learned that on August 17, Robert Fulton and his partner Robert R. Livingston had inaugurated a steamboat service on that waterway between Albany and New York. If Fulton could operate a steamboat service on the Hudson River, Molson reasoned, why should he not do so on the St. Lawrence?

Montreal Brewer John Molson (1763-1836) introduced steamships to the St. Lawrence. (C-115899, LAC)

The notorious St. Mary's Current, caused by the narrowing of the St. Lawrence River as it flows through the Port of Montreal, frequently forced sailing ships to lie at anchor for several days, or even weeks, waiting for a favourable wind. If Molson were to build some steamboats, he thought, not only would he be able to provide a steamboat service between Montreal and Quebec but also a steamboat could tow a becalmed sailing ship against St. Mary's Current.

A year and a half later, Molson with two associates had had the plans for the hull and upper works of a paddle steamer drawn up, and a contract for her construction had been signed. She was to be built at David Munn's shipyard, a short distance upstream from Molson's brewery, and George Platt & Ezekiel Cutter had begun work on her machinery at their foundry in Montreal. Her shafts and many parts for her engines were to be cast and fabricated at the Forges du Saint-Maurice in Trois-Rivières, Quebec.

At a simple launching ceremony on August 19, 1809, the paddle steamer was named *Accommodation*, following which

Molson's Paddle Steamer *Accommodation* is launched in 1809. (*Cape Breton Post*, LAC)

she was towed to the brewery to receive her 6-hp engine and paddlewheels. Her length, according to her register, was 85 feet and her width 16.

After several trial runs to Boucherville and the necessary modifications having been made to her boilers to improve her speed, the paddle steamer *Accommodation* was ready for service between Montreal and Quebec. She left on her maiden voyage on November 1, 1809, and claimed her place in history as the first steamboat in service on the St. Lawrence River.

How it was that the men responsible for her construction, the brewer John Molson, the shipbuilder John Bruce and the mechanic John Jackson, initially got together, we do not know. What we do know is that John Molson funded the project, paying all expenses for the steamboat's construction, as laid down in a formal agreement signed by the three men in June 1809. This included his partners' wages of 7 shillings and 6 pence per day. Those of the Quebec shipbuilder John Goudie for 19¾ days of expertise at the same rate would be agreed upon in a separate contract.

In addition to the method of funding, the agreement specified that when the steamboat was put into service, Bruce would serve as captain and Jackson as engineer, for which each would receive 10 shillings per day, and be given their meals. They would, however, pay for their beer themselves. It was also stipulated that profits and losses would be shared equally by the three men

There was considerable scepticism on the part of the townspeople, and only ten of her twenty berths had been reserved when she left Montreal for Quebec on her maiden trip. Besides the captain and engineer, her crew of six included the river pilot Amable Laviolette, who acted as steward, and two men who served as both stokers and deck hands.

The *Quebec Mercury* put a bold face on her difficult journey when she arrived in Quebec:

> On Saturday morning, at eight o'clock, arrived here, from Montreal, being her first trip, the steam boat, Accommodation, with ten passengers. This is the first vessel of the kind that ever appeared in this harbour. She is continually crowded by visitors. She left Montreal on Wednesday, at two o'clock, so that her passage was sixty-six hours; thirty of which she was at anchor. She arrived at Three Rivers in twenty-four hours. She has, at present, births (sic) for twenty passengers; which, next year, will be considerably augmented. No wind or tide can stop her. She has 75 feet keel, and 85 feet on deck. The price for a passage up is nine dollars, and eight down, the vessel supplying provisions. The great advantage attending a vessel so constructed is that a passage may be calculated on a degree of certainty, in point of time; which cannot be the case with any vessel propelled by sail only. The steamboat receives her impulse from an open, double-spoked, perpendicular wheel, on each side, without any circular band or rim. To the end of each double spoke is fixed a square board, which enters the water, and by the rotary motion of the wheel acts like a paddle. The wheels are put and kept in motion by steam, operating within the vessel. A mast is to be fixed in her, for the purpose of using a sail when the wind is favorable, which will occasionally accelerate her headway.

The claim that no wind or tide could stop her was sadly tried on her return to Montreal, where she required a team of oxen to tow her past St. Mary's Current. Yet, three of her passengers wrote glowingly of the trip, stating in a letter to the editor of a local newspaper that the sole reason for her lack of power was a miscalculation in the amount of fuel, i.e. the firewood that she required. The other aspects of their trip, the food, accommodation and service, they declared, were excellent. She made a second trip on November 15, which completed her first season in service.

Improvements were made to the *Accommodation*'s machinery during the winter, and she began the 1810 season on June 5 with increased horsepower, an improved boiler and paddle-wheel mechanism, two masts, so that auxiliary sails could be hoisted, a camboose (kitchen), and many more berths. Thus enhanced, she continued her regular voyages until October.

At the end of the second season, expenses for the steamboat amounted to £4,000, and receipts to less than £500. Though so far, a financial failure, Molson realized the steamboat's potential and bought out his two partners, who were unwilling to take any further risks.

Molson's next step was to visit Robert Fulton in New York, who convinced him to obtain his engines from the Boulton & Watt works in Soho, Birmingham, in future, and he promptly arranged to go to England to see them. Fulton also convinced him to apply for the sole right to run steamboats in the province, and Molson drafted a petition to that effect. Put forward by Joseph Papineau and Denis-Benjamin Viger, it received the assent of the Legislative Assembly in 1811, but was refused by the Legislative Council.

Fulton's third suggestion that Molson enter into a partnership with him for steam navigation on the St. Lawrence, he declined.

Molson spent the winter of 1810 to 1811 in England, where he visited the Boulton and Watt works in Soho and ordered two steam engines. They were to be delivered in the spring of 1812 accompanied by John Bennet, a 21-year-old engineer who would sign a three-year contract with Molson on his arrival. Bennet was to set up the engines and show Molson's employees how to maintain and run them, as well as teach them the work of a millwright, for a total salary of $400 a year.

The PS *Swiftsure,* the new steamboat in which one of the 30-hp engines was installed, played a part in the War of 1812 moving troops and their supplies, but not the part that Molson had envisaged. He had thought that the government would decide to refit her as a transport and charter her for the entire season. In fact, that only happened from time to time. Shipbuilders on the St. Lawrence were soon building steamboats with far more powerful engines, and their presence on the river had become commonplace.

It was time for steam ferryboats to make their appearance, and the first was the double-ended paddle steamer *Lauzon.* She was built by John Goudie at his shipyard in Quebec's suburb of St. Roch for both passenger and freight service between Point Levy and Quebec. The company that owned her was made up of the prominent Quebec merchants John Caldwell, John Davidson, François Languedoc, Hiram Nicholas, John White, and Goudie himself.

The second steam ferry was the double-ended paddle steamer *Lauzon* built in Saint-Roch, a suburb of Quebec City. (*Quebec Mercury*)

The service, which began in 1818, was much appreciated by the public who crossed on her more cheaply and more comfortably than they had done previously, in a canoe.

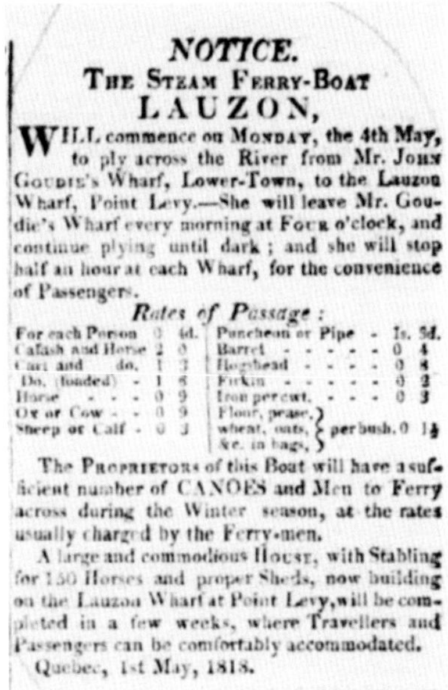

NOTICE.

THE STEAM FERRY-BOAT

LAUZON,

WILL commence on MONDAY, the 4th May, to ply across the River from Mr. JOHN GOUDIE'S Wharf, Lower-Town, to the Lauzon Wharf, Point Levy.—She will leave Mr. Goudie's Wharf every morning at FOUR o'clock, and continue plying until dark; and she will stop half an hour at each Wharf, for the convenience of Passengers.

Rates of Passage:

For each Person	0 4d.	Puncheon or Pipe	-	1s. 3d.
Calash and Horse	2 0	Barrel	-	0 4
Cart and do.	1 3	Hogshead	-	0 8
Do. (loaded) -	1 8	Firkin	-	0 2
Horse - - -	0 9	Iron per cwt.	-	0 3
Ox or Cow - -	0 9	Flour, pease, wheat, oats, &c. in bags,	per bush.	0 1½
Sheep or Calf -	0 3			

The PROPRIETORS of this Boat will have a sufficient number of CANOES and Men to Ferry across during the Winter season, at the rates usually charged by the Ferry-men.

A large and commodious HOUSE, with Stabling for 150 Horses and proper Sheds, now building on the Lauzon Wharf at Point Levy, will be completed in a few weeks, where Travellers and Passengers can be comfortably accommodated.

Quebec, 1st May, 1818.

It cost less to take the ferry than the canoes run by local ferrymen. (*Quebec Mercury*, 1 May 1818)

Local ferrymen, however, accused Goudie of stealing their livelihood and dubbed her "*la chienne d'invention anglaise.*" But the *Lauzon's* captain was chosen from among them and soon other ferrymen had become proud steam ferryboat captains, too.

John Goudie was not content with building just one steamboat, and, the following year he launched the Paddle Steamer *Quebec,* the first of the Quebec-built steamers put into service between Quebec and Montreal that competed with those of the Montreal line. The *Quebec Mercury* of May 11, 1819, wrote:

> The Paddle Steamer Quebec of 1819.
>
> LAUNCH The steam-boat Quebec was launched from Mr. Goudie's Yard, in St. Roch, on Saturday morning. She went off in the best style to the music of the band of the 76th Regt. And the cheers of thousands assembled on that occasion. She proceeded almost immediately down the St. Charles, to Mr. Goudie's wharf in the Lower town. From which she steamed up and down the St. Lawrence during the greatest part of the day, which was spent on board by the owners, and a number of their friends, in conviviality and festivity.
>
> We understand the Quebec will be ready to proceed to Montreal in about a week. Her workmanship, appearance and accommodations, certainly do honour to the builder and the proprietors, and we have no doubt she will greatly add to the comfort of the communications between the two cities. We believe she is the eighth steamboat on the river, six of which have been built in Montreal and only two in Quebec.

By then, steam navigation had captured the public imagination. Each spring, as the ice left the river, the return of the steamboats from their winter quarters in Sorel was eagerly anticipated. On May 12, 1815, the *Quebec Gazette* wrote:

> On Saturday last our harbour presented a gay and animated spectacle, in the arrival of none less a number than eight Steam Boats with flags flying and guns firing, from their winter quarters in William Henry [Sorel]—the *Swiftsure, Montreal, Lady Sherbrooke, Caledonia, Malsham, Quebec, Telegraph* and *Car of Commerce.*

Quebec City Shipbuider John Goudie, 1775-1824. (Courtesy Peggy Bolenbaugh)

At times, however, sport took over from transport, impromptu races enlivening the trip. Egged on by enthusiastic passengers, over-stoked boilers, and tied-down escape valves led on occasion to explosions, not without some fatal results.

Apart from carrying passengers and freight, the steamers undertook all kinds of towing including restraining ships at their launching, and soon the first purpose-built steam tug on the St. Lawrence was put into service.

Known then as a towboat, the 100 ton tug *Hercules* was built by Alexander Young in Munn's Montreal shipyard and was launched in December 1822. Soon she was towing large sailing ships up and down river and past St. Mary's Current, ships that might otherwise have had to lie at anchor waiting for a favourable wind.

In 1824, after visiting Upper and Lower Canada, Edward Allen Talbot wrote:

> There are no less than seven steamboats which constantly ply on the St. Lawrence between Quebec and Montreal, five of which are nearly as large as a 40-gun frigate. They are fitted up in a very elegant manner for the accommodation of the passengers. On each side of their cabins, some of which are large enough to accommodate one hundred persons, there are two rows of berths, one above the other. The berths are supplied with excellent bedding and running curtains. Separate from the gents' cabin is one in each boat for the ladies; in which however they only sleep, for they take breakfast, dinner and tea, in the common room with the gentlemen. Every possible attention is paid to passengers on board of these boats Servants of every description are always in waiting; and tables are daily laid out, exhibiting all the delicacies of the season, and every luxury which this fruitful country affords. In a word, their accommodations of every kind are not at all inferior to those which are to be met with in the most respectable hotels in Europe. (Talbot, *Five Years Residence in Canada*)

That same year, when the 3690 ton raft-ship *Columbus,* followed in 1825 by the 5294 ton *Baron of Renfrew*, each one the largest ship afloat in its time, set sail from Quebec for England, the *Hercules* towed them the 400 km downriver as far as Bic, leaving them there to proceed on their trans-

Quebec from Point Levi. Drawn by James Gray; Aquatinted by J. Gleadah; Published by Willett and Blanford, 1828. (3FDE4C5C, Toronto Reference Library.)

atlantic course. The raft-ships were no more of a challenge to the *Hercules* than the strings of three barges or even three tall ships that she, and later other tugs, could be seen towing on the St. Lawrence River.

By 1825, the citizens of Quebec, like those of many towns and villages all over the world, had become accustomed to the sound of the steamboats' strident whistles on their rivers and lakes. In many places steamboats were making coastal trips as well as long distance sea trips in stages. Moreover, a paddle steamer was about to be laid down in the Port of Quebec that would make a quantum leap, a leap destined to make history.

CHAPTER 2

The Quebec and Halifax Steam Navigation Company

In 1830, the Canada we know today did not exist. The provinces of Nova Scotia, New Brunswick, Lower Canada, and Upper Canada, which with Prince Edward Island made up British North America, lay side by side but went their separate ways. It was clear that their union would strengthen them economically, socially, and defensively, and many suggestions had been made over the years as to how that might be accomplished.

No one was more convinced of the need for union than Richard John Uniacke, the Attorney General of Nova Scotia. He strongly believed that the five provinces that made up British North America should consist of only two, one of Lower and Upper Canada united, and the other, of the united three Maritime Provinces. He delivered a memoir to that effect to the Colonial Office in London in 1806, which pointed out the many benefits that would result. It led only to discussion, yet his was one of several similar proposals that reached London from the provinces around that time.

As the years passed and Uniacke persisted in his campaign, he was supported by others.

> In 1822 a plan for the union of Upper and Lower Canada presented to the House of Commons in London caused a political stir in the colony. In Montreal some eminent businessmen, Molson among them, formed a committee in support of the bill which held a public meeting and collected 1,452 signatures. (Alfred Dubuc, "John Molson," *Dictionary of Canadian Biography*, vol. VII)

And that same year, an article in a Halifax newspaper, copied in the *Montreal Gazette*, reported the suggestion that a steamboat service between Halifax and Quebec be established to "carry the mail and improve communications

Richard John Uniacke, Attorney General of Nova Scotia, petitioned London in favour of the union of British North American provinces. (Painted 1811 by Robert Field. Courtesy Uniacke Estate Museum Park. Photo: Nova Scotia Museum)

between the two provinces." Steamboats had been navigating successfully on the St. Lawrence River since 1809, and it was proposed that one of them be purchased for the new service by a company to be formed in the two towns.

> The first cost of the vessel together with the outfits are estimated at £10,000 and it is intended that there should be 200 shares at £100 each; it appears that the passage from one port to the other would occupy six days, and therefore the vessel would call at Pictou and take in coals, and complete her voyage round in twenty-four days, and thus make 14 trips during the summer season. (*Montreal Gazette,* 30 July 1822.)

Although the idea did not immediately bear fruit, when London was struck in 1824 by what some referred to as a "steamboat mania," many were ready to put their money down. As a result, *The Canadian Magazine and Literary Repository* was able to report in May that a company had been floated in London with capital of £50,000 under the name, The Canada and Nova Scotia Steam Navigation Company. Far from being an isolated or far-fetched idea, and apart from the expected benefits to the British North American colonies, it was regarded as an extension to the transatlantic steam service that was being planned between Ireland and Halifax by another newly formed company, the American and Colonial Steam Navigation Company.

Though it was made public that the Canada and Nova Scotia Steam Navigation Company would begin service the following spring, it was not to be. Like many other ventures that were floated at the time, the company failed to survive the stock market crash in December. In fact, it was one of 67 out of the 70 newly formed steam navigation companies that no longer existed in 1827. As for the American and Colonial Steam Navigation Company, it suffered too, despite such illustrious patrons as the Duke of York and the Duke of Wellington, but was able to re-structure and limp on.

The governments of both Lower Canada and Nova Scotia had done their part. In February of 1825 Lower Canada appointed a committee to "consider the expediency of promoting communication between the ports of Quebec and Halifax by means of a steam vessel or vessels." Appearing before the committee, none other than Justice Norman Fitzgerald Uniacke, the eldest son of Richard John Uniacke of Halifax, spoke enthusiastically of the many benefits that would result. As far as trade was concerned, he argued, Nova Scotia had fish, plaster of Paris (gypsum plaster) and coal, which Lower Canada needed, while Lower Canada produced wheat and bread stuffs, which Nova Scotia then obtained from the United States.

He himself had travelled from Quebec to Halifax, he said, taking 21 days to go and 23 to return, a tedious voyage by sail due to the light winds. He felt confident that by shortening the trip, a steamboat service would attract a considerable number of patrons and bring together the people of the four provinces.

Many Americans, he added, who currently visited the Canadas, would travel from Upper to Lower Canada, Gaspé and the Bay of Chaleur if there were a steamboat. It could stop at Charlottetown, Pictou, and Sydney, and refuel at Cape Breton, where there were coalfields. Steamboats, he concluded, were now running around the coast of England, and the technical know-how could easily be obtained from there.

But that would not be necessary, for Charles Wood, the Scottish naval architect and pioneer steamboat builder, was engaged at the time in building the raft-ship *Baron of Renfrew* on the Île d'Orléans, in the St. Lawrence River off Quebec. Both he and his right-hand man, shipbuilder Joseph Ritchie, were summoned to appear before the Lower Canada parliamentary committee enquiring into the projected steam navigation service.

In answer to the committee's questions, Charles Wood replied that navigation could be performed by steam vessels with perfect safety. That a steamboat of 500 tons burthen should be employed, and that such a vessel could carry 200 tons of freight and sixty passengers, and cost from ten to twelve thousand pounds. Such a steamboat could also tow vessels between Quebec and Halifax. As to the most efficient means of encouraging such an undertaking, he considered that not less than a £1,000 premium should be offered.

Mr. Ritchie was of the same opinion. Moreover, following his testimony he laid before the committee the plan of a

Halifax from Dartmouth, 1838. Engr. Bartlett.

steam vessel built by Charles Wood, which was then plying between Leith and London impelled by two 50-hp engines. The engines, he said, were contracted by Messieurs Watt and Boulton, of Soho, Birmingham. The vessel was fitted for the accommodation of one hundred and fifty passengers, and generally made her passage in about one hundred and fifty hours, a distance of about five hundred miles.

Their positive testimony and that of several others led to a bill being passed in the Provincial Parliament of Lower Canada in March 1825. It offered a subsidy of £1,500 sterling in three equal annual payments to the first person or company who built or had built a steam vessel of not less than 500 tons and navigated her between the ports of Quebec and Halifax during four years. Payments of £500 would be made after the first voyage of each of the three first years.

Meanwhile, the House of Assembly of Nova Scotia had also taken up the matter. On April 2 that same year, it resolved "that a sum of two hundred and fifty pounds be granted and placed annually, for the next three years, at the disposal of his Excellency the Lieutenant Governor, to enable him to cooperate with the Government of Lower Canada in the establishment of Steam Boats to run between Quebec and Halifax." It was ratified by the Council three days later.

But all this would prove of no avail. False claims made by many steamboat owners in the past had disillusioned the public, and it took three years for the merchants to regain confidence in the project.

In 1828, the report of the Quebec Committee of Trade recommended that navigation by a steamboat be established between Quebec and Halifax and that an application to government for two packet boats per month between Halifax and Falmouth, England, during the summer season be made.

This led to a new act being passed in the House of Assembly of Lower Canada, which doubled the amount of subsidy offered to £3,000. Payments of £1,250, £1,000, and £750 respectively were to be made at the close of each of the first three seasons, contingent on the vessel being kept insured for not less than the sum to be advanced at the end of the season and the amount already advanced. The steamboat was to be navigated "during such part of the year as the navigation between the said Ports shall remain safe and open, the dangers of the navigation always excepted," and it could enter intermediate ports at the discretion of the owners.

Once again, Nova Scotia followed the lead of her sister province, and increased her offer. Early in March, a resolution moved by Richard J. Uniacke to the effect that a committee be formed to study the steamboat question passed the Assembly with a two-thirds majority. The three-man committee, with Uniacke as chairman, acted quickly issuing a report recommending that the sum of £1,500 be granted over a period of three years towards the establishment of a line of steamboats to ply between Halifax and Quebec.

It was accepted by both the House of Assembly and the Legislative Council. That is to say, Nova Scotia, too, doubled the subsidy offered.

The increase in both subsidies, added to improved market conditions, had the desired effect. Within a few months, The Quebec and Halifax Steam Navigation Company had been incorporated with capital of £16,000 at £25 per share. According to a report of a meeting of the shareholders published in the *Acadian Reporter* in June, 204 shares were owned in Quebec, 135 in Montreal, 91 in Miramichi, and 139 in Halifax. The 235 shareholders are listed in Appendix F.

A committee was then formed in each of the two provinces. The Committee of Management of the Lower Canada Committee comprised Quebec's leading merchants: William Finlay, Jeremiah Leaycraft, Joseph Stowe Shaw, William Walker, John Caldwell, William Price, John Leather, John Saxton Campbell and Charles Felix Aylwin. Prominent on the Nova Scotia committee were the three Cunard brothers, Henry, Joseph, and Samuel, of which the latter held the powers of general management and financial control. Samuel was also elected the company's agent in Halifax, while Joseph became the agent in Miramichi, New Brunswick.

The business community in Halifax did not receive the venture with the same degree of enthusiasm as did that of Quebec. Despite the number of shares sold in Halifax, it was reported at a meeting of the mercantile body that the project had been coldly received. It was "generally felt that it would be preferable to put the effort into increasing trade with the West Indies and Europe." The public, however, felt otherwise.

CHAPTER 3

The Team

The Quebec committee had charge of the construction of the vessel and included many prominent local shipbuilders and businessmen. By July the management committee had drawn up specifications for the hull and on August 1 specifications for the engines were exhibited in the company's committee room in Quebec. A month later the contract for the steamboat's hull and spars was awarded to the lowest bidder, the shipbuilder George Black.

Black was considered a first-class shipbuilder. He had built twenty-four large sailing vessels since settling in Quebec following the War of 1812. In partnership with the timber merchant and shipowner John Saxton Campbell since 1825, he had earned an enviable reputation for the quality of the building and repairing carried out under his management at the Campbell & Black shipyard at Cape Cove. Moreover, the partners had such confidence in the steamboat venture that they each guaranteed that the steamboat would be launched by the last day of April, in eight months' time, under a £1,000 penalty for non-compliance and £50 for each day it was late, pledging all their individual property present and future as security. Furthermore, additional security was pledged by the sailmaker and shareholder James Hunt.

The contract for the engines was awarded to Bennet and Henderson, a company set up by the two men following Bennet's three years of service with Molson. The partners agreed to make the entire machinery, consisting of two side-beam engines with cylinders each of 50 inches in diameter, two iron boilers, and all other parts, with the exception of the crankshafts, that were to be ordered from Robert Napier in Scotland, and to deliver the steamboat within two months of the hull's arrival at St. Mary's Foundry, their premises in Montreal. As the date set for that to happen was the end of April, she was therefore to be delivered by the end of June 1831.

The Scottish engineer John Lowe, recently hired by Bennet and Henderson, was to design and make the plans for the 90-hp engines and the other parts of the machinery. An experienced engineer, he had previously designed steam

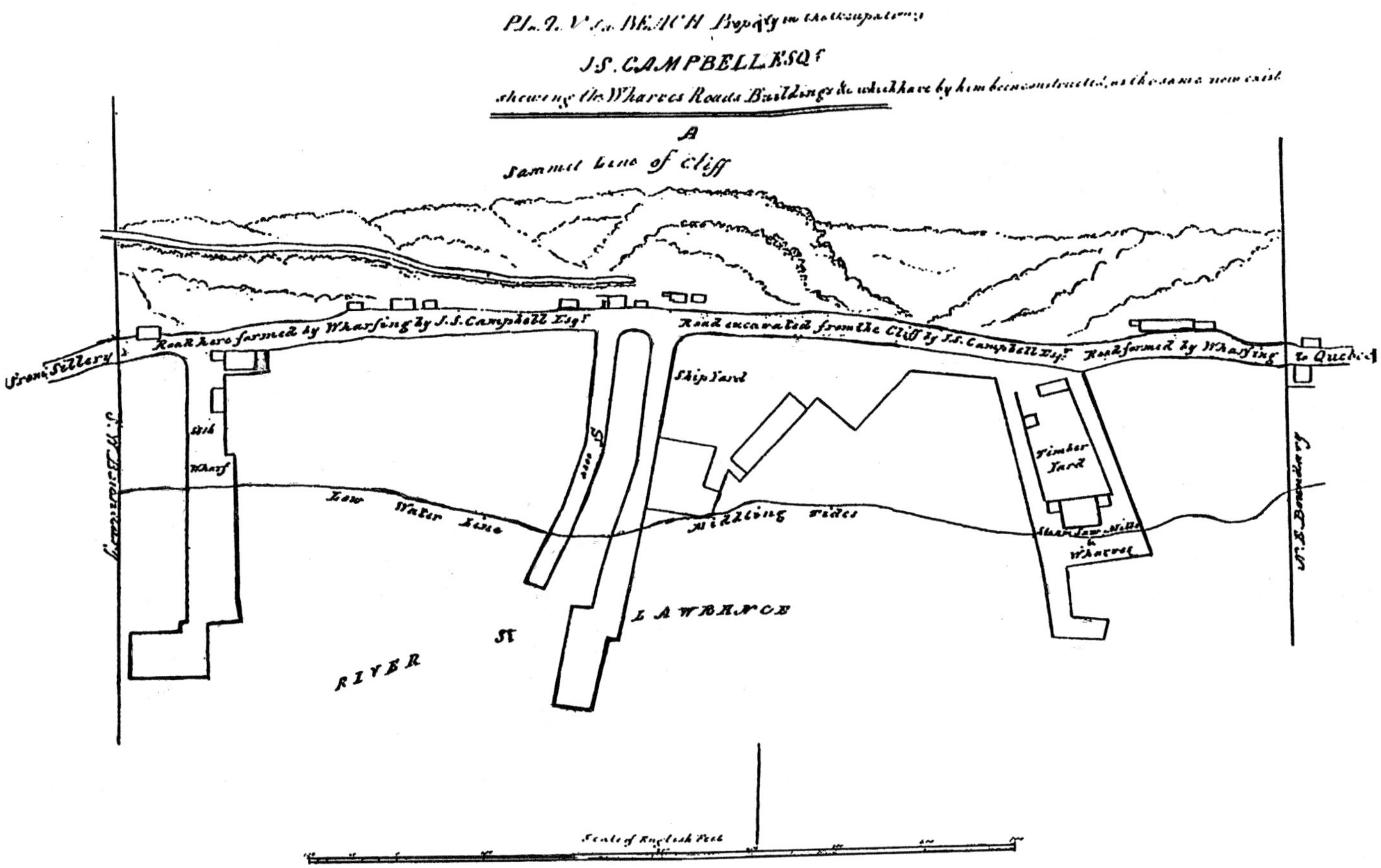

Plan of the Cape Cove shipyard. John Adams, 1831. (Attached to deed of sale of sawmill, right, Campbell to Hall, 25 Jan. 1831, gr. F.-X. Garneau, BANQ)

The Campbell & Black shipyard at Cape Cove, with its white mould loft (centre) and floating dock alongside. (Neg. 103102, LAC)

John Lowe (1779-1876), native of Dunkeld, Scotland, designed the *Royal William*'s engines. (Courtesy, Fred L. Colby, Jr.)

James Goudie, who superintended the construction of the *Royal William*. (Courtesy Peggy Bolenbaugh)

engines for mills, boats, and other applications both in the British Isles and in France.

The contract to superintend the construction of the steamboat was given to James Goudie, the eldest son of the late John Goudie. James had served his time at the Simons shipyard in Greenock, Scotland, where he had acquired the skills and knowledge of a competent shipbuilder. Soon after his arrival there, he had been named assistant foreman to William Simons himself, and in his last year was chosen foreman apprentice for the building of a steam yacht for Dom Pedro of Portugal and Brazil. He recounts in a letter that in the spring of 1830, with his apprenticeship drawing to an end:

> I was told that Mr Simons, wanted to see me, and hurried to his office. "James," he said, "I have been asked to send a capable man to Quebec to superintend the construction of a paddle steamer like those we have built here recently. You have worked on all of them, and I'm sure you will do well. You can take a copy of the plans, and adapt them over there to their requirements." I was taken aback. "I am only twenty, sir," I said. "Do you not think I am too young to handle a job like that?" "If I did not think you capable, I would not recommend you," Mr Simons replied.

So in May of 1830, following his passage across the Atlantic and a 750-mile journey up the St. Lawrence River, James found himself stepping off the gang plank of the ship that brought him home to Quebec with a half-model and a roll of steamboat plans.

He would later write that he was well supported by his assistant, Thomas Oliver, who had apprenticed with Black,

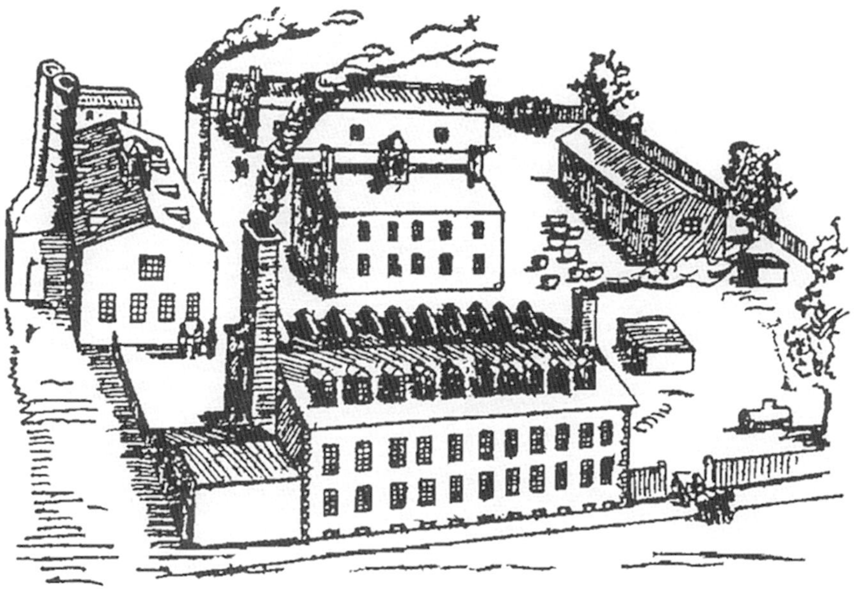

Bennet and Henderson's Foundry complex that housed St. Mary's Foundry, from the *Daily Witness*, 20 March 1897.

John Bennet, whose foundry made the engines for the *Royal William*.

and that Thomas' 13-year-old brother Edward was one of the other apprentices on the job. Both would play an important part in Quebec's shipbuilding industry later on. Thomas became the most prolific 19th century builder of sailing ships in the port of Quebec. Edward, who moved on from shipbuilding in Quebec to ship brokerage in Liverpool, might have done as well had he not misjudged the market following the Crimean War and urged shipbuilders to build when the sailing ship market was about to collapse.

We learn from other sources that the experienced Maltese shipbuilder Louis Shickluna, who began his career when he was eleven, working beside his father and grandfather in the royal naval dockyard there, and later would set up his own successful shipyard in St. Catharine's, Ontario, served as foreman of the construction, and spent the 1831 season as the *Royal William*'s mate.

Finally, Captain John Jones a highly regarded navigator in the Royal Navy, was given the steamboat's command.

Thomas Hamilton Oliver, James Goudie's young assistant, who became Quebec's most prolific 19th century shipbuilder. (Photo: G. Marcil)

Louis Shickluna, a young shipwright from Malta, served as foreman. He later became an important shipbuilder in St. Catharine's, Ontario. (Courtesy, Donald Holmes)

What were his credentials? The local newspaper reported, on his arrival from England in 1830, that he had been "engaged" by Admiral Ogle in 1829 in determining the latitudes and longitudes of the North American coasts and the Gulf and River St. Lawrence. In fact, he had been in charge of the survey and renewal of all Admiralty maps and charts for that area and had, besides, set up a new naval hydrographic station for the Royal Navy in Halifax.

The steamboat company was now ready to sign the agreement with the government, and did so on September 3, 1830.

The keel of the steamboat had been laid the previous day.

CHAPTER 4

Building the Steamboat

WITH THE PLANS OF SEVERAL STEAMERS that he brought from Scotland to guide him, James Goudie designed the new vessel to meet the specifications drawn up by the Quebec and Halifax Steam Navigation Company, allowing for the different properties of Canadian timber. He then laid down the lines on the mould loft floor.

According to the specifications, the hull was to be built in very much the same way that the hulls of sailing ships were built, except that it was to be wide enough to house a large paddlewheel on each side. It was to have a rock elm keel, with keelson, transom and fashion pieces of oak, all fastened with oak treenails. The cabin half-deck and forecastle deck were to be made of 3-inch-thick yellow pine 4 to 6 inches wide. It was to have oak plank around the combings and where extra strength was needed, and all to be copper fastened. There were to be two metal pumps and a brass-mounted mahogany steering wheel. The bottom was to be painted green, while the masts, spars and yards were either to be painted or given a coat of bright varnish. A 22-ft skiff, a 19-ft jolly boat, and two other service boats, all built of oak, were to be supplied.

Evidence of the care with which the steamboat was built is to be found in George Black's letter-book, in which construction details for all twenty-four ships laid down by Black since his arrival in Quebec are listed, the dimensions of each timber and spar carefully noted to the nearest quarter of an inch! The book is unique in Quebec, where no other 19th-century shipbuilder's book used for this purpose is known to exist. Moreover, George Black's signed copy of C. F. Partington's *Shipbuilders' Complete Guide,* published in 1826, is in the collection of the library of the Quebec National Assembly.

The specifications for the hull and rigging of the steamboat can be found in Appendix A; those for the machinery, though not the oak paddlewheels, in Appendix B.

The Campbell & Black mould loft on whose floor James Goudie laid down the lines of the paddle steamer *Royal William*. (Detail, l'Anse au foulon. A1. BANQ)

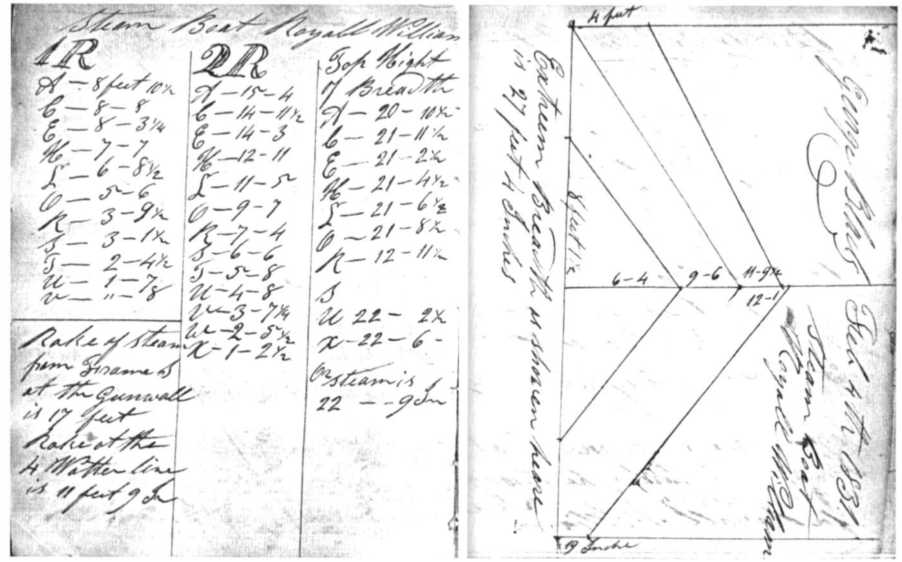

One of the nine pages of George Black's Letter Book, in which dimensions of the timbers of the *Royal William* and other information regarding her construction are given. (AP-P-203, BANQ)

With timber coves on both sides of the Campbell & Black shipyard and a steam sawmill immediately adjacent to it, Black would have had no problem obtaining the timber he required. Knees were brought directly to the shipyard from farms in the countryside around Quebec. Names of the farmers and details of where they lived are also noted in George Black's letter-book.

Work progressed on the hull of the steamboat through the winter, and when she was *in frame*, the plankers began

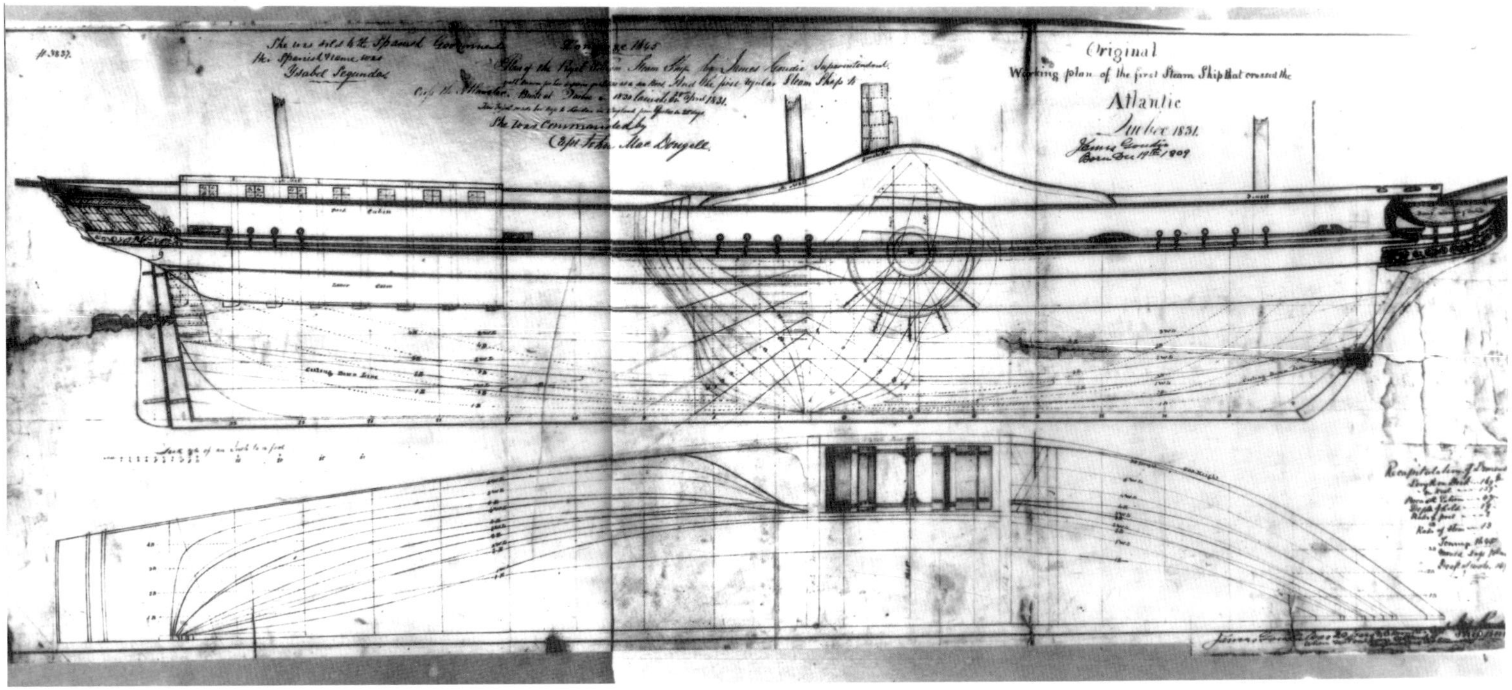

Plan of the *Royal William*, drawn by James Goudie. (Collection of the Chicago Historical Society)

planking the exterior, the interior, and part of the deck. The rest of the deck would be planked in Montreal after the engine had been fitted. As caulkers filled the last seam in the planking, a scroll head and quarter badges were added, and the steamer was ready for her launching.

Friday, April 29, 1831, the day set for the event, was declared a public holiday and a day of celebration. Several of the ships in port lent their bunting, and thousands of spectators gathered on every vantage point above and around the shipyard. The band of the 32nd Regiment stood ready to mark the event in appropriate manner aboard the PS *Richelieu,* while at the Citadel a gun salute was prepared. The Governor General and his wife arrived from an earlier launching at a nearby shipyard, and Lady Aylmer, as godmother to the vessel, struck the bow with a bottle of Madeira decorated with a garland of flowers as His Excellency pronounced the customary words "God bless

The Cape Cove shipyard, Friday afternoon, 29 April, 1831. The hull of the *Royal William* lying alongside the mould loft, the steamboat *Richelieu*, with spectators aboard (right centre). The Canada Floating Dock on the near side of the crowd-filled quay. (Watercolour, James Pattison Cockburn, 20-1-139, neg. C 12469, LAC)

the *Royal William* and all those who sail in her." The roar of the guns and the crashing strains of the regimental band mingled with the hurrahs of the crowd as the steamboat slid slowly into the river.

It was reported that the shipbuilders Allison Davie and George Taylor were among the shareholders of the Steam Navigation Company who attended the launching. Other shipbuilders, who were shareholders, such as Hyppolite Dubord, William Lampson, and John Munn, are not likely to have missed the occasion. The economy was once again in the doldrums; expectations for the success of the venture were high.

Little time was wasted before the *Royal William* was sent to Montreal to be completed. The towboat *British America* left her at the foot of St. Mary's Current, opposite the Bennet and Henderson foundry, during the night May 2. There were now two months in which the machinery was to be installed, part of the deck planked over, and the interior finished and furnished. So far the execution of the contract was on time.

The contract for the interior joinery of the steamboat had been given to the carpenter and joiner, Joseph Andrew, of Montreal, and was to be completed within two months of the vessel's arrival there. Andrew was to finish the interior with the best quality pine, with certain exceptions such as the staircase, whose steps were to be of oak, and balustrades and handrail, of mahogany. The full specifications are in Appendix C.

The cabinetmaker and upholsterer Charles Fry was to furnish the steamboat with bedding of the best quality, mahogany tables, side tables and sideboards, worsted damask curtains, and rush-bottomed sofas and chairs. (See Appendix D.)

The ladies' cabin in the *Charlevoix*, en route from Quebec to Montreal, by Katherine Jane Ellice. (LAC, Acc. No. 1990-215-22)

The St. Lawrence River steamboats were known for the care with which they were furnished, and the ladies' cabin on the *Charlevoix* is likely to have been finished and furnished by Joseph Andrew and Charles Fry who received the contracts for the steamboat *Royal William*.

Bennet & Henderson were allowed two months to make and install the engines, while the two other contracts were

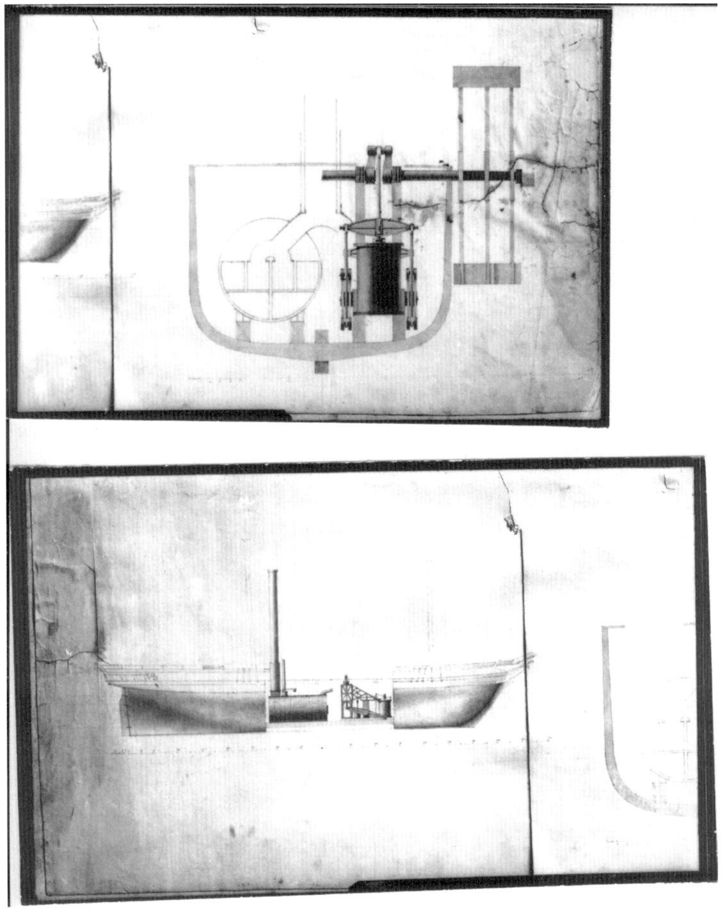

Original drawings of the machinery of the *Royal William* in the collection of the Dossin Maritime Museum. (Courtesy, Fred L. Colby, Jr.)

completed. The steamboat therefore was to be ready by the end of June the following year. In retrospect, the time allowed for the three contracts seems unrealistic, and we know that the work had not been completed one week after the due delivery date, on July 10. It was on that day, that the *Montreal Gazette* reported that in changing berths, "she [the *Royal William*] touched on a rock, which swung her bows round to the full force of the current, her anchor dragged and the vessel grounded on a bank." And there she stayed for eleven days despite all attempts to move her. It was not until August 13 that the *Royal William* eventually left Montreal.

After calling at Sorel and Trois-Rivières, she arrived at Quebec the next evening and tied up at the King's wharf, "competent judges" in Montreal having declared before she left that she was "in no way inferior to the finest steam packets in the old world."

CHAPTER 5

On the Saint Lawrence River

TEN DAYS LATER, ON AUGUST 24, 1831, the *Royal William* left Quebec on her first voyage to Halifax, with John Jones, RN, as Captain; James Sutherland, William Peddle, and Thomas Cleghorn, 1st, 2nd and 3rd Officers respectively; James Laing, Chief Engineer; and Robert Bartley, James George Danter, and James Harris, 2nd, 3rd and 4th Engineers.

Stops were scheduled at Chatham, Douglastown, and Newcastle on the Miramichi River, and at Charlottetown and Pictou, both on the way there and on her return. She carried 20 cabin and 50 steerage passengers, some freight, and 120 tons of coal.

The *Acadian Recorder* of Saturday, September 3, described her reception in Halifax:

> The day of her arrival was very bright; and on Intelligence of her arrival being communicated from the signal station, considerable interest was exhibited by the inhabitants to obtain a view of their visitor. Some sought the house tops, some the wharves and others the surface of the harbor in boats to gratify their curiosity. At 10 o'clock she appeared off Mauger's Beach, and keeping near the western shore, proceeded up the harbor in most majestic style. Several guns and enthusiastic cheers welcomed her arrival, and she answered in a similar manner. Her beautiful fast shining appearance, the powerful and graceful manner in which her paddles seemed to pace along, and the admiral's command which her helmsman had over her, afforded a triumphant specimen of what steamships are.

John Fitzgerald, the Halifax pilot who picked her up, received a fee of eight dollars for his part in her safe docking. Though only 23 years old at the time, he had the necessary experience. He had already brought in many of the warships that visited Halifax, and would later bring in the first Cunarder that visited the port. According to J. G. Danter, the *Royal William*'s second engineer, Samuel Cunard boarded her on several occasions while she lay at the Cunard wharf. He asked many questions, Danter wrote, taking careful notes of the answers, and was particularly interested in her sailing qualities, her speed and her consumption of coal.

No doubt Cunard was already among those who were seriously considering establishing a transatlantic steamship line.

When the *Royal William* called at Charlottetown on her return trip, she elicited the following account in the *Royal Gazette,* September 13:

> THE ROYAL WILLIAM STEAMSHIP — On Wednesday, this elegant and substantial vessel touched here on her way from Halifax to Quebec via Miramichi. Her arrival was greeted with firing of cannons, and the cheers of the numerous spectators, whom the novelty of the sight had attracted to the wharves and other convenient sites for viewing her approach, as she dashed into our beautiful harbour against wind and tide. She had hardly dropped her anchor before she was surrounded with boats, filled with young and old, all eager to gratify their curiosity by inspecting her interior arrangements; and it is but justice, both to officers and men to say that all who went on board, and they were not a few, were received with every mark of civility. After remaining about four hours, she again got under weigh for Miramichi – where she arrived safely next morning. … Her accommodations are of the first description. Her cabins are elegant and the sleeping berths, of which there are about 50, admirable. The round house contains a spacious dining-room, handsomely fitted up capable of accommodating 100 persons. The steerage also is roomy and comfortable, and there is ample space on deck.

The steamer was back in Quebec on September 13, the return having taken two days longer because she was delayed for four days at her ports of call and by fog.

After four days in port at Quebec, she left on her second trip to Halifax, which took from October 18 to 26. She arrived back in Quebec November 9, having faced a hurricane on her way between Halifax and Pictou. A letter from John Lowe, the designer of her engines, to his parents describes the journey.

> I have had a letter of the success of the *Royal William* Steam Ship's second trip between Quebec and Halifax in which on her return from Halifax to Quebec she encountered one of the heaviest gales of wind experienced for a long time in this quarter. The Capt. who has been at sea for 25 years, says he never saw such a heavy gale except once when *HMS Acorn* was lost, yet the *Royal William* weathered the gale and returned to Quebec in good order not even as much as a bolt about her engines was broke although both her quarters were knocked away and a part of her deck by the weight of the sea which struck her, and highly gratified as the company were by her performance of her first trip, they are particularly so by her success and stability on her second, which is no small satisfaction to me being the first steamboat that ever went to sea in America and my first management in Canada.

In the report of her journey under gale conditions that was published in the October 28 edition of the *Acadian Recorder*, the captain stated that when the wind freshened, he directed his course seaward, and on the morning of the gale found himself about 40 miles to windward of White Head. The Steamer was *laid to* under her mizzen; her paddles moving slowly to prevent lee way and she thus rode out

The City of Quebec, Taken from the Harbour. Day & Hague, Litho.; Joseph Bouchette, 1831. (C9EC8178 ,Toronto Reference Library)

the hurricane with open hatches and without shipping a single sea. Moreover, he added that he, the crew and the passengers were all of the opinion that the *Royal William* was to be preferred in a gale to either King's ship or merchantman.

High praise from an officer of the Royal Navy!

The steamer's third and last trip to Halifax that season took 21 days overall.

Sorel 1839: The church, at whose doors the *Royal William* was sold, and the wide beach on which she and other steamboats wintered. (Watercolour, Philip John Bainbragge. C002469, LAC)

But the enthusiasm with which the steamboat was received in each port of call did not translate into profit; the company was losing money and changes had to be made. The intention to employ her in the West Indies or on an Atlantic crossing during the winter had had to be abandoned. The Company's charter did not allow for travel beyond Halifax, and it was feared that she might lose the government subsidies should either be attempted. So, like many of the other St. Lawrence River steamboats, the *Royal William* wintered 170 kilometres upriver from Quebec, in Sorel.

Plans were made to cut down expenses the next season. A petition was presented to the House of Assembly of Lower Canada in November, which stated that in causing the steamboat to be built, the Company had believed that it was in the interest of the three provinces of New Brunswick, Nova Scotia and Lower Canada; that the Company had incurred much expense and some risk on the part of the voyage of the steamboat from Cape Canso to Halifax and back; and that the purpose of building the *Royal William* would be equally attained were the steamboat not to sail beyond the Port of Pictou in Nova Scotia. Cited, in particular, was the heavy cost of insurance beyond that point.

As a result, a bill was prepared to amend the Act, which passed into law in February of 1832. In future, the voyages of the *Royal William* would end at Pictou. This cannot have pleased Samuel Cunard, who thereby lost his fees as agent in Halifax, fees that the committee in Quebec had considered too high. A second request, that the government purchase the 68 unsold shares of the company was turned down. They had already been advertised for sale unsuccessfully.

With the less onerous journey, the *Royal William* might have had a different career had a cholera epidemic not been brought to Quebec from Europe. On her first trip in 1832, the steamboat was refused entry into Miramichi when she arrived there, having lost one of the crew to cholera and with several others ill. Held in quarantine for 23 days, she proceeded then to Halifax only to be quarantined again. The trip from Quebec to Halifax and back took a total of 53 days, and she was unable to continue her service that season. Once again, she would winter in Sorel but before she did so, her owners would send her on a humanitarian mission.

The 301 ton brig *Acadia* bringing 320 emigrants had run ashore on Green Island (Isle verte), 125 miles to the east of Quebec. When the offer of the *Royal William*'s agent to send her to the rescue of the passengers and crew at cost was turned down due to the lack of the necessary funds to pay the bill, the Management Committee of the Quebec and Halifax Steam Navigation Company stepped in and waived the fee.

The steamboat was dispatched and returned to Quebec carrying 160 of the travellers, grateful to have escaped the harsh winter conditions. Twenty-seven others had died, most having succumbed to an epidemic of measles, while the rest had been taken ashore on the south shore opposite the island.

It would appear that proper care was not taken of the steamboat that winter. In a letter of May 7, 1833 to Sir Rupert George, Nova Scotia's provincial secretary, Samuel Cunard, bitterly blamed the Quebec committee for allowing the frost to burst the pipes and injure the machinery. "I do not think the (Provincial) bounty should be paid in consequence of the Boats only making one trip during the whole Season," he wrote. "They have already received £3,975 from this Province the whole of which is lost by the Management of the Quebec Committee and the object in view frustrated."

Blame was also laid on others, including Captain Jones. He was accused of gross extravagance by James Goudie, who wrote of his expensive manner of conducting affairs, having everything done in man-of-war fashion. Heavily in debt, the company had no option but to obtain a mortgage loan of £5,000, the maximum allowed by its charter.

But the mortgage was not enough to meet the bills that kept accumulating. On April 3, 1833, at the doors of the parish church in Sorel, the *Royal William* was sold at auction to the mortgagees for the amount of the mortgage. Two weeks later, the six new owners, James Forsyth, Jeremiah Leaycraft, Henry Lemesurier, Matthew Bell, Noah Freer, and Henry John Caldwell, all Quebec merchants, offered to transfer the purchase to the company's shareholders if they agreed to cover the purchase price and all other expenses before June 1. But that did not happen.

The *Royal William,* under the new command of Captain John McDougall, the former Captain of the Trinity House lightship *Brilliant,* was back in Quebec May 8, 1833. She was no longer the property of the Quebec and Halifax Steam Navigation Company, no longer the mail carrier between the two ports. During the rest of the month she undertook excursions to Grosse Île on Thursdays and Fridays, and was available for other service on the river.

On June 4, she set sail for Boston, where it was hoped that a purchaser would be found. She carried 155 passengers and made stops at Gaspé, Pictou, and Halifax on the way. She was the first British steamer to visit Boston, and was received with great enthusiasm. A salvo of guns welcomed her on her arrival, and a band played *God Save the King* as she steamed into port, while her many visitors included the U.S. President, Andrew Jackson. She spent nine days in Boston, but no purchaser was found.

That otherwise successful trip led to the owners' decision to send her to England for sale. Soon an advertisement appeared in the newspapers offering passage to London for £20, exclusive of the wine bill.

Thus, the *Royal William* left Quebec on August 5, 1833, with four cabin passengers, Mr. Clark of Upper Canada, Rev. Sweeny and daughter of Jamaica, and Mr. Caszier. She stopped at Pictou, where she coaled, had her engines and her boilers overhauled, and waited for the rest of her passengers, who arrived late from Prince Edward Island.

She left Pictou for London August 18 with just seven passengers and a crew of 36; Captain John McDougall in command; Jas Douglas, Jas. Geo. Danter, and J. Short, her 1st, 2nd, and 3rd engineers. Her cargo consisted of six spars, one box, one trunk, some household furniture, a harp, and a box of stuffed birds. Almost all her hold, however, was filled with the fuel she would need for the voyage, 254 caldrons of Cape Breton coal.

The "box of stuffed birds," Robert Ker tells us, "was a collection made by the Rev. Thomas McCulloch, the local Presbyterian Minister and naturalist." Others say that they were the outcome of John James Audubon's recent expedition to the Lower Saint Lawrence.

No doubt the owners of the *Royal William* felt that they were doing the right thing in sending her to England, as they were unable to operate her at a profit in Canada and were unable to sell her in Boston. There were many, however who, like the *Pictou Colonial Patriot's* editor, saw her leave with regret.

> "When the idea of connecting Quebec and this province by steam was first started," he wrote, "we hailed the opening prosperity with a very long article, pointing out the advantages which would ensue, and though our predictions have not been realized, we have never seen cause to change our opinion."

Apart from the losses that the steamboat company's shareholders suffered, not only did it bring an end to the

"The *'Royal William'* pioneer of transatlantic steam navigation, leaving Pictou for London one hundred years ago."
(Print. Adam Sherrif Scott. Ink and coloured ink on paper. M16738 McCord Museum)

many advantages of a steamboat service between the provinces at the time, but despite the blame being attributed to the cholera epidemic, Canadian pride was hurt and the finger pointing was intense

Samuel Cunard, for instance, argued that the Quebec Committee had acted irresponsibly. The Nova Scotia Legislative Assembly agreed, and the final £500 of the Nova Scotia subsidy was not paid to the Steamboat Company. Instead, it was to be transferred to the local Steamboat Company in return for an improvement in its ferry service, though eventually it went towards the upkeep of the province's roads.

Regardless of his opinion of the Quebec committee, the *Royal William* is generally believed to be the catalyst that led Cunard to set sail for England (in 1839), where he tendered successfully to build a line of transatlantic mail steamers, steamers that would become the Cunard Steamship Line.

CHAPTER 6

Crossing the Atlantic

NO DOUBT the *Royal William*'s owners were disappointed to find only four passengers from Quebec and three from Prince Edward Island who were willing to risk the Atlantic crossing. Perhaps the public was afraid of the rough Atlantic seas straining the paddle steamer's machinery in bad weather. Moreover, those aboard soon experienced that situation, for on reaching the Newfoundland banks, a violent storm arose snapping the foretopmast. The starboard engine failed, and so much water was shipped that the chief engineer urged the captain to put into port, convinced that the steamer would founder. MacDougall, however, had confidence in his ship. The pumps were manned, and she ran on one engine while her engineers repaired the damage to the other

But that was not the only problem with which they would have to contend. Every four days, time was lost while the steamer laid to for 24 to 26 hours while the scale formed by the saltwater feed was chipped off the inside of her boilers. Moreover, another week was spent overall repairing the damage for which that procedure was no doubt partly responsible.

Despite the erratic voyage, the *Royal William* made the crossing in 21 days, overcoming the major problems faced by sea-going steamers at the time. Not only was her heavy load of coal sufficient to complete the journey, but the regular removal of the encrustation from her boilers, which was necessary before the surface condenser for marine engines was developed, prevented it from becoming dangerous. Furthermore, steam was not blown off in order to lengthen the periods between de-scaling, thus avoiding the waste of both steam and coal.

Safely reaching the English Channel on September 9, the *Royal William* stopped at Portsmouth to allow her passengers to disembark, before proceeding to a shipyard on the Isle of Cowes. There, the necessary upkeep and repairs were carried out, and the crew gave the steamer's exterior a fresh coat of paint—they had painted the interior on the way over.

Thomas White's Shipyard in West Cowes, Isle of Wight, 1834, where the *Royal William* received maintenance and was painted before carrying on to London "in fine style." (White's Archive)

"Cowes. September 9. Arrived the *Royal William* steamer from Quebec and Pictou. Sailed 19th ult." *The Times*, London.

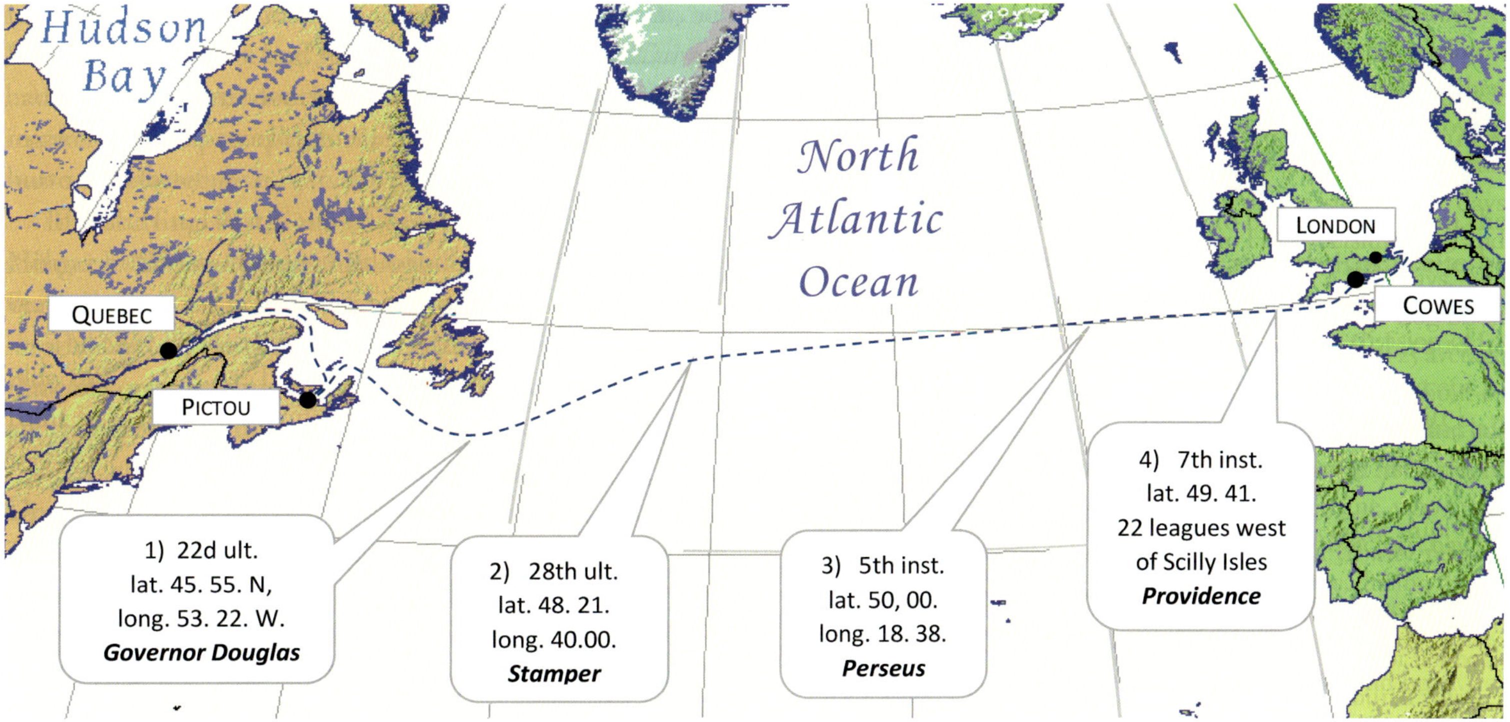

"COWES, Sept. 9th. Arrived ... ..., and the steamer *Royal William* of Quebec, McDOUGAL, master, from Pictou in 21 days, for London. The master reports that the Mining Association steamer had arrived at Sidney, about the 8th ult., and the barque Economist had arrrived at Pictou, from Leith, on the 12th ult., all well. and spoke with the following vessels, namely:- (1) On the 22d ult., in lat. 45. 55. N, long. 53. 22. W., the *Governor Douglas*, of Miramichi, from Cork, out 16 days, for Quebec; (2) 28th ult., in lat. 48. 21., long. 40.00., the *Stamper*, of Maryport, SCURR, master, from Cardigan, out 21 days, for Quebec; (3) 5th inst., in lat. 50, 00., long. 18. 38., the *Perseus*, of London, from Liverpool, out 11 days, for Miramichi; and on the (4) 7th inst., in lat. 49. 41., 22 leagues west of Scilly, the *Providence*, of and for St. John's, Newfoundland, from London, all well." *Morning Post*, London, September 11 1833 .
(BASE MAP SOURCE: Public domain maps of the Atlantic Ocean http://ian.macky.net/pat/license.html accessed March 17, 2017)

Readers of the *London Gazette* learned of her arrival and that the Despatches she brought over had been forwarded to the Admiralty, Treasury, and War and Ordnance offices. The names of vessels spoken to on the voyage appeared in London's *Morning Post* the following week and soon spread through the country, from London to Edinburgh, and from Canterbury to Dorset.

The *Royal William*'s arrival in Cowes was reported in the same way in local newspapers as that of any other vessel arriving at a port in Britain. The reports were then copied by distant others. Some wrote that she had faced a storm, but no unusual details were given of the crossing. There was no mention that she had been sent to England for sale, and certainly no claim of her being the first vessel to steam the entire way across.

For one thing, even well-educated members of the public had been misled by the ongoing persistent false claims made of the *Savannah*'s crossing in 1818 and believed that a steam crossing the entire way had been achieved long before.

And for another, as the esteemed popular scientist Dr. Dionysius Lardner had made it clear that it was impossible to build a steamer that could cross the Atlantic without re-coaling, who would believe that the *Royal William* had succeeded in doing so? "You might as well talk of making a journey from New York or Liverpool to the moon," he had said. That being understood, the arrival of a steamboat from across the Atlantic did not raise an eyebrow.

Two months later, however, when the report of her safe arrival had reached Canadian and U.S. papers, the *Royal William*'s owners no doubt appreciated an article carried by both the *Washington Intelligencer* and the *New York Courier & Enquirer,* with the comment "As it is in contemplation very shortly to run a Steam packet from New York to London or Liverpool, we deem the following highly interesting and satisfactory."

BY THIS DAY'S POST.

DEAL, Sept. 9.—ENE.—Arrived from the river and sailed, the Mary Ann, Hornblow, for India; the British Tar, ... for Quebec; and the Atwick, for New South Wales. ... the homeward bound, the Chapman, the Barossa, the ... Brougham, and an American line ship.

FALMOUTH, Sept. 8.—N. fresh.—Arrived, the Harlequin, Hunter, from Gonaives.

LIVERPOOL, Sept. 8.—Arrived, the Grecian, Whitby, and the Claremont, Stevens, from Jamaica; the Grace, Tumian, from Savanella; the Effort, Gibson, from Halifax; the Re... former, Littlejohn, from Teneriffe; and the Roselle, Davidson, from Smyrna. Off port, the Lady H. Ellice, from Virginia. Sailed, the Frances Ann, Hay, for Bengal.

COWES, Sept. 9.—Arrived, the Royal William steamer from Quebec and Pictou, sailed ... ult.

The *Royal William* steam-packet, with passengers, arrived at Portsmouth on Monday evening, from America, having made her passage between land in seventeen days, during which time she encountered a strong gale on the Great Bank, and contrary wind with the exception of six days. Her despatches have been forwarded to the Admiralty, Treasury, War, and Ordnance Offices.

DEAL, SEPTEMBER 9.

Sept. 12.—... for the River, the Barbadoes ... from Barbadoes; John Stewart, Rimington, from Demerara; Thames, Cameron, from Sierra Leone; Royal William, McDougal, and Ann, Long, from Quebec; Elizabeth and Ann, Jeffery, from Teneriffe; Flora, Levigne; New Phœnix, ...rton; Millicent, Cousins; Mandingo, Bergin; Hibernia, ...okny; St. Leonard, Rutherford; and West Indian, Simpson, from Jamaica.

COWES, SEPT. 9.

Arrived the American brig *Arcturas*, FOQUE master, from St. Jago de Cuba in 85 days, for orders; the English brig *Fanny*, IRWIN master, from Havannah in 62 days, for orders; and the steamer *Royal William*, of Quebec, M'DOUGAL master, from Pictou in 21 days, for London. The master reports that the Mining Association steamer had arrived at Sidney about the 8th ult., and the barque *Economist* had arrived at Pictou, from Leith, on the 12th ult., all well, and spoke with the following vessels, namely:—On the 22d ult., in lat. 45. 55. N., long. 53. 22. W., the *Governor Douglas*, of Miramichi, from Cork, out 16 days, for Quebec; 28th ult., in lat. 48. 21., long. 40. 00., the *Stumper*, of Maryport, SCURR master, from Cardigan, out 21 days, for Quebec; 5th inst., in lat. 50. 00., long. 18. 38., the *Perseus*, of London, from Liverpool, out 11 days, for Miramichi; and on the 7th inst., in lat. 49. 41., 22 leagues west of Scilly, the *Providence*, of and for St. John's, Newfoundland, from London, all well. Sailed the French Government steamer *Sphinx*, Capt. SARLAT, for Cherbourg; the English barque *Arcadia*, ANDERSON master, for Antwerp.

Newspapers that reported on the *Royal William*'s arrival in England include the *London Standard*, September 10, 1833, the *London Gazette*, September 10, the *Morning Post*, September 11, the *Berkshire Chronicle*, September 14, *Western Flying Post*, September 16, *Kentish Gazette Deal*, September 17.

CHAPTER 7

Portugal

Consigned to the London bankers Geo. Wildes & Co. for sale, the *Royal William* was turned over by them to the shipbrokers, Wilcox & Anderson. Had some other firm been involved, the rest of her story would no doubt have been very different, for the partners Brodie McGhie Wilcox & Arthur Anderson concentrated on the Spanish and Portuguese trade. In fact, it was said of them that "they shipped anything from guns to giraffes between Britain and the Iberian Peninsula."

They wasted no time in selling the *Royal William* for £10,000 sterling, to the prominent London shipowner, Joseph Somes, and then arranged her charter to the Portuguese government, expecting that that country would decide to buy the steamer from Somes when the charter ran out.

Portugal was in turmoil. When Napoleon's army crossed its border in 1807, the Portuguese royal family and their court, together with members of the government, civil servants, judges, and wealthy merchants, including their domestic servants, had escaped to Brazil. They set up the capital of the Portuguese Empire in Rio de Janeiro as part of *The United Kingdom of Portugal, Brazil and the Algarves,* and they thus avoided deposal by Napoleon, the fate of the German and Italian royal families.

Joined later by other countrymen, it is estimated that at least 10,000 people moved at that time from Portugal to Brazil.

> On November 29, 1807, a day before the French army entered the city, Dom João and his Spanish Borbón queen, Dona Carlota, fled by the only route available to them: the sea. The scene rivalled the fall of Saigon. A convoy of three dozen frigates, brigantines, sloops, corvettes and ships-of-the-line, with the entire Portuguese court on board, 10,000-strong, set sail for Brazil with a British escort, braving the winter storms of the Atlantic. Religious dignitaries, government ministers, military officers, their families and servants—and Dom João's mad mother, Queen Maria—made up the complement of passengers on the three dozen vessels. The ships also carried the imperial regalia, the royal carriage, a piano and several tons of books and documents, the paperwork of empire. (Patrick Wilcken, *Empire Adrift*)

Antigo Palácio dos Vice-reis, atual Paço Imperial, no Rio de Janeiro
The Imperial Palace, built in the 18th century as a residence for the governors of colonial Brazil. From 1808, it was used as a royal palace by King John VI of Portugal, as King of Portugal, and later as King of Brazil. Correspondence of the Marquês do Lavradio.

Brazil would remain the capital of the Portuguese Empire until 1821, when King John and his court returned to Portugal to deal with the Liberal Revolution that had broken out the previous year. The King's 22-year-old son Dom Pedro was left in Brazil, as regent. The following year, however, dissent in both countries led Dom Pedro to declare the independence of Brazil, of which he was proclaimed Emperor.

Unwilling to leave Brazil, when his father died five years later, Pedro renounced the Portuguese throne in favour of his minor daughter Maria da Gloria, and named his younger brother Miguel, Prince Regent. However, when he heard that Miguel had usurped the throne, he abdicated the Brazilian throne to his son, Pedro II, and returned to Portugal, and in the course of six years of civil war, with his Spanish supporters and the help of Britain and France, Pedro fought to regain the throne for his daughter.

Though a long-time ally of the Portuguese, Britain was unwilling to send her regular army to take part in the war, but the Foreign Enlistment Act was suspended and she allowed Portugal to recruit an auxiliary legion of 10,000 mercenaries. Thus, with a British crew commanded by Captain McDougall, the *Royal William* left Portsmouth on January 4, 1834, on her first voyage under charter to the legitimate Portuguese government. She was carrying a boatload of mostly Belgian mercenaries to Portugal to fight on Maria's behalf, and exchanged them for invalids and disbanded soldiers that she brought back to London.

The *Royal William* was then laid up off Deptford victualling office and remained there until orders were received to fit her out to run between Oporto and Lisbon and, after a voyage between the two ports, she was sent to Cadiz to fetch a cargo of specie.

King Pedro of Portugal abdicates in 1824 in favour of his daughter, Princess Maria da Gloria. (Artist, Domingos Antonio de Sequeira, 1826 Museu nacional de Arte Antiga.)

The steamer was still sailing under the British flag, for her proposed sale to Portugal had been blocked by Admiral Charles Napier, Commander of the Portuguese fleet. Though a firm believer in steam navigation, Napier was of the opinion that naval vessels should have iron rather than wooden hulls.

Thus, in September 1834 the *Royal William* was lying in Lisbon awaiting new orders, after completing her last charter to the Portuguese.

CHAPTER 8

Spain

MEANWHILE, neighbouring Spain was undergoing similar succession problems to those of Portugal. King Ferdinand VII had died in September 1833, and his brother Don Carlos and his supporters were contesting Ferdinand's young daughter Isabel II's right to the throne. Moreover, Spain was still reeling from the cost of the Napoleonic invasion, which the loss of Spain's newly independent American colonies had compounded. The country could ill afford another debilitating war.

However, several European countries had acquired steamboats that served as speedy auxiliary vessels in their fleets, and though the Spanish treasury was badly depleted, the government decided that Spain should do the same. The matter of whether to purchase a suitable steamer in Bordeaux or have one built in Spain was still under consideration when an explosion occurred in a frigate undergoing repairs in the Spanish naval dockyard at Ferrol.

The high cost of repairing the dock led at first to the steamboat project being postponed, but when it was learned that Carlos, the Pretender, was preparing an expedition in Holland, that he had acquired a steamer, had had two ships loaded with munitions, and that he intended to land mercenaries and armament in a Basque port in Gupuzcoa on the north coast of Spain, the priorities of the legitimate Spanish government changed.

There was no time to build a steamer. Asked to enquire what steamboats were available in London, the Marquess of Miraflores, the Spanish ambassador there, acted quickly. Learning that Portugal had rejected the offer to buy her, he found a guarantor for the asking price of £10,000 sterling for the *Royal William*, and arranged for her purchase through the Spanish ambassador in Lisbon where the steamboat was berthed. The sale was completed on September 10, 1834, and, at Miraflores' urging, Captain MacDougall and the entire British crew were engaged. The *Royal William* was temporarily re-named *Isabel*, which was changed on her arrival in Spain to *Isabel Segunda* or *Isabel II.*

The *Isabel Segunda* became part of a squadron of the Spanish navy named the British Auxiliary Steam Squadron, of which Commodore Frederick Henry was appointed

Isabel II. Dámaso Cabo y Rochina de Castro. *Historia de Cabrera*. Madrid 1845. (Museo del Prado, Madrid)

Commander-in-chief. Captain McDougall was given the rank of Commander in the Spanish Navy and was guaranteed a salary of £600, as well as a contract to supply the squadron with provisions from Lisbon.

As might be expected, Spanish naval officers were outraged to hear that a British captain, British engineers, a British crew and British marines, had been engaged to man the *Isabel Segunda,* and would, moreover, receive higher pay than theirs, but Ambassador Miraflores held his ground. He argued that the officers and men of the Spanish navy had no experience in steam navigation, that operating a steamboat required special training, and that it was not the time for an inexperienced crew to be in charge, adding that the *Isabel Segunda* would be working independently of other ships in the Spanish Navy, in the British squadron, or Escuadron Auxiliar Britanico de buques de vapor.

To the Spanish Government's enquiry as to what would have to be done to upgrade the *Isabel Segunda* to a naval vessel, Wilcox replied that no essential repairs were needed, as she was only three years old and had been solidly built with copper fastenings. She had recently undergone a refit, he wrote, and had been re-sheathed with copper at a Thames shipyard. However, her roundhouse should be removed and her upper deck examined, as it might have to be replaced. Her gunwales would need strengthening to support her artillery, and her beams also would need extra support. Her engines were in good condition, and she carried replacements for them. Her boilers, not surprisingly, would require replacing in a matter of a couple of months.

As to armament, he continued, she would require two 32-pounder guns for midships, six carronades of the same calibre, and some rockets with the necessary launching apparatus. Lodging would have to be built for the officers and crew. The oak paddlewheels were highly vulnerable to enemy fire, and should be changed for iron wheels, and a capstan to raise the anchors should be installed. Coal bins must be provided as well as extra sails. Lookout yards and additional cordage would be needed, and hammocks for the crew, with rails from which to hang them. Finally, storerooms would be required for the rockets and gunpowder.

Commodore Henry, Wilcox added, found the *Isabel Segunda* to be an excellent ship, and stated that with those few improvements she would make a good naval vessel. While undergoing refit, Wilcox concluded, a second steamer would replace her.

The *Isabel Segunda* left Lisbon and sailed for the north coast of Spain, where Don Carlos' troops were entrenched and proceeded to take part in the blockade. Meanwhile, the *City of Edinburgh,* which was to take *Isabel Segunda*'s place when the latter returned to England for refit, was undergoing conversion to a man-of-war.

The plan would change, however, when the *Isabel Segunda*'s boilers gave out sooner than was expected and she was withdrawn from service. The *City of Edinburgh* was not

The paddle steamer *Isabel Segunda,* ex. *Royal William* of Quebec, off Gravesend, Kent. She is wearing the Red Ensign of the British Merchant Navy, while waiting the delivery of a Spanish flag, having passed her own national ensign on to the PS *Reina Gobernadora*. Painted by Stephen Dadd Skillet, 1834. Exhibited at the Royal Academy, London. Bought and presented to the Literary & Historical Society of Quebec by F. Lascelles. Now in the collection of the Musée national des beaux arts du Québec (L74.260). Photograph: Patrick Altman.

yet ready to replace her, so a third British steamer, the *Royal Tar,* was chartered to serve in the squadron during her refit.

When the *Isabel Segunda* ex *Royal William* arrived back in England for her conversion, Commodore Henry, Captain MacDougall, and her crew of 160 men were transferred to the newly acquired PS *Royal Tar,* which left immediately for Spain. Arriving off the Cantabrian coast on November 29, the *Royal Tar*'s name was changed to *Reina Gobernadora.* She proceeded to patrol with the squadron between San Sebastian and Bayona, where arms were being landed, and to support the troops of the British Legion who were facing a formidable enemy in the embattled area.

Don Carlos had the support of the Pope, the more extreme elements of the Catholic Church, the Conservatives, and a large number of Basques in northern Spain. Furthermore, a steady supply of arms was reaching them from France, across the Bay. Incensed that foreigners, heretics as they saw them, had taken sides and were opposing them in a domestic war, the Carlists offered no quarter.

Meanwhile, at the Sheerness Dockyard in England, the *Isabel Segunda* received the outfit she required as a man-of-war. Her frame was strengthened, and her engine room was fitted with three new boilers and a surface condenser, the recent invention of Samuel Hall. The surface condenser would allow her boilers to be fed with fresh distilled water, and consequently would no longer become encrusted with minerals that had to be removed every few days.

The *Isabel Segunda* then took General Alava, the Spanish Ambassador to London, and General Evans, commander of the Auxiliary Legion, with most of his staff, to St. Andero (Santander). Sailing on to San Sebastian, she joined the patrol blockading the coast between that port and Fuente Arabia, on the Franco-Spanish border.

In 1837, when the contracts of the British crew expired, the *Isabel Segunda* returned to London, where the crew was paid off and Captain John McDougall parted with the ship with which he had made history, thus ending Canada's part in the career of the PS *Royal William/Isabel Segunda.*

Her history as the *Isabel Segunda,* however, would carry on. Back in Spain, with her new Spanish crew, she was credited with capturing ten of the fifteen fishing boats of a flotilla that was running the blockade.

In 1839, the blockade finally ended, The *Isabel Segunda* was sent to Bordeaux for an overhaul and it was decided to hulk her, build a new hull and sell the hulk to the French. The new steamer though smaller, was fitted with the *Isabel Segunda*'s Canadian engines, her armament, and the parts of her rigging and superstructure that were sound. Resplendent with a newly carved bust figurehead of Queen Isabel II, the now two-masted steamer *Isabel II* then took up service in the Spanish Mediterranean Coastguard fleet.

Ten years later the *Isabel II*'s name was changed to *Santa Isabel,* when a larger and more powerful steamer, built in England, was given her name. As the *Santa Isabel,* she

The steamers *Phoenix* and *Salamander* of the British Squadron bombard the Carlists in Ayote and the Fort of Lugariz, while the people of San Sebastian applaud from the quays opposite. (Unknown artist. Colección del Untzi Museoa-Museo Naval.)

El vapor *Isabel II* en la marejada. The Ill-fated *Santa Isabel* grounds in the Bay of Gibraltar. (Oil Antonio Brugada, Empresua: Central Librera Ferrol S.L.

performed a variety of duties, from carrying or escorting dignitaries, to combat roles. In 1859, she took part in the bombardment of Larache during the Spanish-Moroccan War, which was destined to be her last time in action. The following January she grounded in the Bay of Gibraltar, where she had taken refuge during a violent storm. But fortunately, the crew, with one exception, reached safety by means of a rope bridge set up to the shore. The *Santa Isabel,* however, was less fortunate and all that remained of the *Royal William* went down in her to the sea floor.

Built to carry mail and improve communications between the provinces of Lower Canada and Nova Scotia, which might have had a successful outcome had there not been the cholera outbreak, the paddle steamer *Royal William* had a very different career. Her destiny was decided by two men who "shipped anything from guns to giraffes between Britain and the Iberian Peninsula."

Her legacy, however, remains her Atlantic crossing.

CHAPTER 9

Honour in her Homeland

IN THE DAYS WHEN A SHIPWRIGHT relied on a wooden half-model of the hull of the vessel he was building, on which to base its lines, several half-models that had served that purpose hung in the mould loft of the Campbell and Black shipyard in Quebec.

In 1875, the yard was sold and Henry Dinning, one of the new owners, presented the half-model of the *Royal William* to the Literary and Historical Society of Quebec.

Proudly displayed on the library wall, the half-model remained there until 1891, when Archibald Campbell, a vice-president of the Society, moved at the Annual General Meeting of the Society that the model be shown at the Royal Naval Exhibition that was to be held in London the following year. The members agreed and a letter was written to Captain Alfred Jephson, the Secretary of the Exhibition Committee, offering to send the model over.

In answer, Jephson thanked the Society, but added that the committee would only accept it, if it could be well authenticated that the *Royal William* was really the first steamship to cross the Atlantic, as the Naval Committee was "very particular about proper authentication."

Far from being surprised, members of the Society were no doubt expecting the condition imposed on the acceptance of their offer, for the validity of the steamer's claim had not been without challenges over the years.

However, the report drawn up by Frederick Wurtele, the librarian of the Society, satisfied the Committee in London, and Captain Jephson wrote back accepting the loan and assuring the Society that the model would be well placed. He was so glad, he wrote, to have something from old Quebec at the Naval Exhibition.

The model was accordingly sent to London, where it was given a place in the Seppings Gallery in Chelsea under the official number 4736.

The exhibition attracted two million visitors in the five months in which it was held, and when it closed, along with a Commemorative Diploma, the model was returned

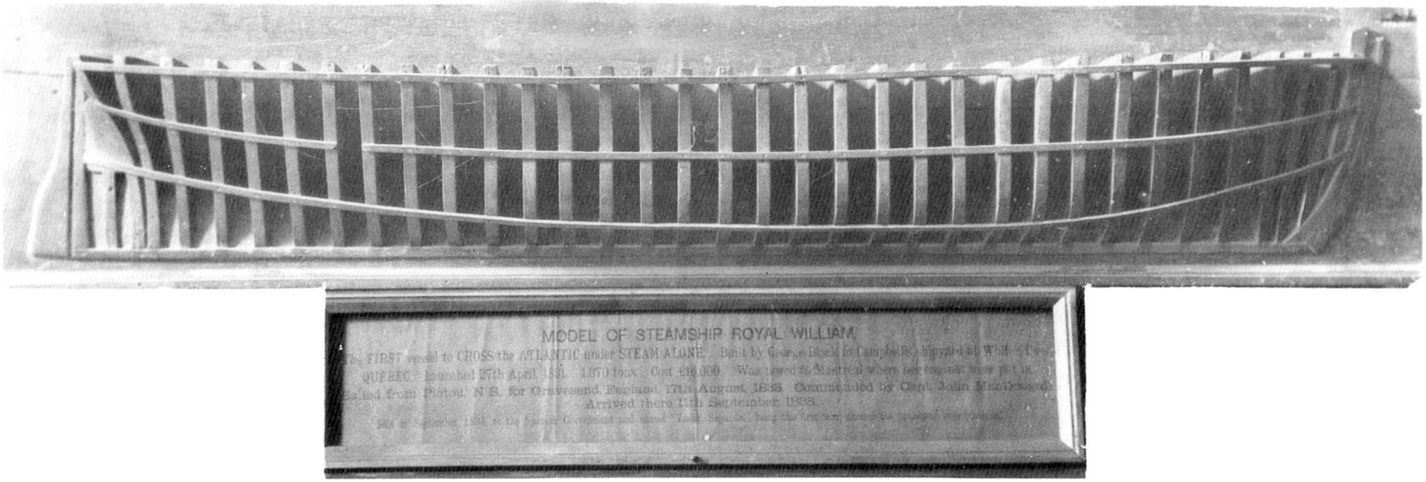

The hawk's or crow's nest 1/2-model of the *Royal William* presented to the Literary and Historical Society of Quebec by Henry Dinning, now at the Musée de la civilization, Quebec.

to the Literary and Historical Society and to its place on the library wall.

Years of controversy were thus swept aside by the exhibition committee in London. It was time that Canada, too, gave the *Royal William* her due.

The renewal of interest in the steamer had come about in 1875 as the result of a letter sent to the *Times* of London by a certain A. M'Donald. It had led to the editor of the Toronto *Daily Globe* asking the prominent Canadian civil engineer Kivas Tully to research the *Royal William*'s history.

Tully complied, and two years later his inaugural address as president of the Engineers' Club of Toronto told the story he had pieced together, under the heading "Transatlantic Steam Navigation; Canada's Claim to its Inauguration." When the *Daily Globe* carried his speech shortly after, both James Goudie, the superintendent for the construction of the steamer, and James Danter, the steamer's 2nd engineer, wrote to the paper adding details to his account.

The *Globe* article would prompt James Goudie to donate a copy of the original working plan for the construction

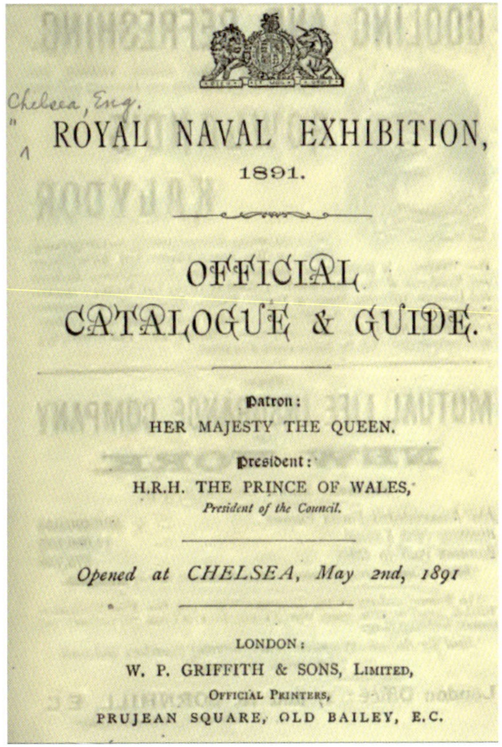

Chelsea, Eng.

ROYAL NAVAL EXHIBITION,

1891.

OFFICIAL
CATALOGUE & GUIDE.

Patron:
HER MAJESTY THE QUEEN.

President:
H.R.H. THE PRINCE OF WALES,
President of the Council.

Opened at CHELSEA, May 2nd, 1891

LONDON:
W. P. GRIFFITH & SONS, LIMITED,
OFFICIAL PRINTERS,
PRUJEAN SQUARE, OLD BAILEY, E.C.

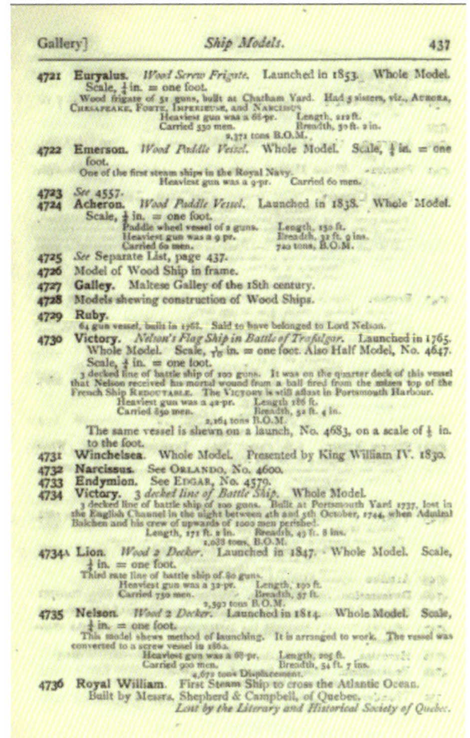

Gallery] *Ship Models.* 437

4721 **Euryalus.** *Wood Screw Frigate.* Launched in 1853. Whole Model. Scale, ¼ in. = one foot.
Wood frigate of 51 guns, built at Chatham Yard. Had 5 sisters, viz., AURORA, CHESAPEAKE, FORTE, IMPERIEUSE, and NARCISSUS.
Heaviest gun was a 68-pr. Length, 212 ft.
Carried 530 men. Breadth, 50 ft. 2 in.
2,371 tons B.O.M.

4722 **Emerson.** *Wood Paddle Vessel.* Whole Model. Scale, ¼ in. = one foot.
One of the first steam ships in the Royal Navy.
Heaviest gun was a 9-pr. Carried 60 men.

4723 *See* 4557.

4724 **Acheron.** *Wood Paddle Vessel.* Launched in 1838. Whole Model. Scale, ¼ in. = one foot.
Paddle wheel vessel of 2 guns. Length, 150 ft.
Heaviest gun was a 9 pr. Breadth, 32 ft. 9 ins.
Carried 60 men. 720 tons, B.O.M.

4725 *See* Separate List, page 437.

4726 Model of Wood Ship in frame.

4727 **Galley.** Maltese Galley of the 18th century.

4728 Models shewing construction of Wood Ships.

4729 **Ruby.**
64 gun vessel, built in 1762. Said to have belonged to Lord Nelson.

4730 **Victory.** *Nelson's Flag Ship in Battle of Trafalgar.* Launched in 1765. Whole Model. Scale, $\frac{1}{10}$ in. = one foot. Also Half Model, No. 4647. Scale, ¼ in. = one foot.
3 decked line of battle ship of 100 guns. It was on the quarter deck of this vessel that Nelson received his mortal wound from a ball fired from the mizen top of the French Ship REDOUTABLE. The VICTORY is still afloat in Portsmouth Harbour.
Heaviest gun was a 42-pr. Length 186 ft.
Carried 850 men. Breadth, 52 ft. 4 in.
2,164 tons B.O.M.
The same vessel is shewn on a launch, No. 4683, on a scale of ⅛ in. to the foot.

4731 **Winchelsea.** Whole Model. Presented by King William IV. 1830.

4732 **Narcissus.** See ORLANDO, No. 4600.

4733 **Endymion.** See EDGAR, No. 4579.

4734 **Victory.** 3 *decked line of Battle Ship.* Whole Model.
3 decked line of battle ship of 100 guns. Built at Portsmouth Yard 1737, lost in the English Channel in the night between 4th and 5th October, 1744, when Admiral Balchen and his crew of upwards of 1000 men perished.
Length, 171 ft. 2 in. Breadth, 43 ft. 8 ins.
1,028 tons, B.O.M.

4734A **Lion.** *Wood 2 Decker.* Launched in 1847. Whole Model. Scale, ¼ in. = one foot.
Third rate line of battle ship of 80 guns.
Heaviest gun was a 32-pr. Length, 190 ft.
Carried 750 men. Breadth, 57 ft.
2,593 tons B.O.M.

4735 **Nelson.** *Wood 2 Decker.* Launched in 1814. Whole Model. Scale, ¼ in. = one foot.
This model shews method of launching. It is arranged to work. The vessel was converted to a screw vessel in 1862.
Heaviest gun was a 68-pr. Length, 205 ft.
Carried 900 men. Breadth, 54 ft. 7 ins.
4,672 tons Displacement.

4736 **Royal William.** First Steam Ship to cross the Atlantic Ocean. Built by Messrs. Shepherd & Campbell, of Quebec.
Lent by the Literary and Historical Society of Quebec.

Royal Naval Exhibition Catalogue. Ship model included on the bottom of the third page above: "4736 *Royal William* First steam ship to cross the Atlantic Ocean. Built by Messrs, Campbell and Black of Quebec." (Literary and Historical Society of Quebec archives, BANQ)

THE FIRST ATLANTIC STEAMSHIP,—Mr. A. M'Donald, 1 Long Acre, writes to the *Times* as follows: —"Sir John Hawkshaw, in his address before the British Association, falls into a common, and hitherto uncorrected, error respecting the first steamship which crossed the Atlantic. Five years before the Sirius and Great Western made their successful attempts to do this, the steamship Royal William sailed from Quebec on the 18th of August, 1833, and after two or three days' detention at Pictou, Nova Scotia, arrived at Gravesend on the 11th of September, thus making the trip in about the same time as that taken by the first Cunard boats to Boston. The Royal William was built at Three Rivers (*sic*), and fitted at St. Mary's Foundry, Montreal, with engines made in this country. So far as my boyish recollection of the vessel serves me, I think she was about 500 tons burden. I remember very well her departure for England, but in order to be sure, I called at Lloyd's some weeks ago and was courteously shown the Register for September 12, 1833, in which I found, under "Gravesend," the announcement of arrivals on the 11th, the following: —"Steamship Royall William, McDougal, Quebec." Several years before a vessel called the Savannah, fitted with an engine and paddles, crossed from Savannah, Georgia, in 31 days. The paddles were removable. Her engines were only used 18 days. When the Savannah entered the Channel off the coast of Ireland, the smoke from her funnel brought down upon her a gun brig detached from the Channel squadron, under the impression that she was a ship on fire. The Savannah was a full-rigged ship, and although she advertised her sailing and for passengers, no one was brave enough to ship aboard of her. As the Savannah was not a steamship, but merely a sailing vessel, with a temporary arrangement for steaming on board, to the Canadian Royal William must be accorded the honour of being the pioneer of our present large Atlantic steam fleet. What became of this vessel subsequently I am uncertain, but have the impression that she was sold to the Portuguese Government." (Letter published in the *Birmingham Daily Post*, August 30, 1875. Note. The *Royal William* was built in Quebec, not Three Rivers.)

of the *Royal William* to the Chicago Historical Society in 1881. Goudie himself had drawn it up fifty years earlier in Quebec, and had taken it to North Evanston when he moved to the US in 1865.

Though Goudie's gift to the society was widely reported, the *Royal William*'s story might have faded into oblivion had Archibald Campbell not proposed that the model be exhibited in London at the Annual General Meeting of the Society in January 1891, and had he not given a speech extolling the steamer's accomplishment at a meeting of the Literary and Historical Society of Quebec two months later. For the report of his speech attracted the attention of a heavyweight in the person of Sandford Fleming, the Scottish born Canadian engineer, best known as the inventor and staunch promoter of world-wide standard time

Convinced after additional research of the importance of the steamer's achievement, Fleming presented his own account of the *Royal William* and her crossing of the Atlantic at a meeting of the Canadian Institute in Toronto on December 17, 1892, the Institute agreed by adopting Fleming's resolution:

> [T]hat suitable measures be taken to establish a memorial tablet in honour of the men associated with the building and sending to sea of the *Royal William*, and that the members for the City of Toronto be requested to obtain permission for the tablet to be placed in a fit position in the Parliament Buildings in Ottawa," and "that it be remitted to the Council to invite the co-operation of societies or individuals and to take such other means as may be needful to carry out in the best manner the spirit of the resolution.

Fleming then persuaded a meeting of the Nova Scotia Historical Society in Halifax to support the initiative of the Canadian Institute to honour the *Royal William* by the following resolution:

> [T]hat this society respectfully seconds the recommendation of the Canadian Institute that the Government of Canada should adopt some means of communicating the fact that the first vessel propelled by steam through the entire voyage across the Atlantic was the '*Royal William*' built at Quebec and owned in Canada, (and) which made the pioneer voyage from Quebec to London in 1833. And further resolved that this society suggests the expediency of having the evidence proving the pioneer ocean steamship to have been a Canadian vessel properly collated, and published in pamphlet form under the auspices of the Government.

The *Journal of the House of Commons* records that it was Sandford Fleming himself who laid the report of the *Royal William*'s crossing before the Joint Committee of the Library of Parliament and saw the project through. This involved obtaining the consent of the Privy Council to a memorandum presented by the Minister of the Marine, at Cabinet meeting on February 21, 1894, and agreed to one week later.

It was decided that in view of the *Royal William*'s accomplishment, the ideal time to hang the Memorial Tablet would be at the Second Colonial Conference to further preferential Trade and Communications among the British Colonies, which was to take place in Ottawa four months later, in June.

Accordingly, on June 28, 1894, following the opening ceremonies of the Colonial Conference, the Governor General of Canada, the Earl of Aberdeen, unveiled a commemorative brass plaque on the wall of the corridor to the library which read:

> "In honour of the men by whose enterprise, courage and skill the *Royal William* the first vessel to cross the Atlantic by steam power was wholly constructed in Canada and navigated to England in 1833 the pioneer of those mighty fleets of

First Prize design, won by the architects, Fuller and Jones, for the Proposed Parliament Buildings in Ottawa. Wood engraving. (*Illustrated London News*, 1859)

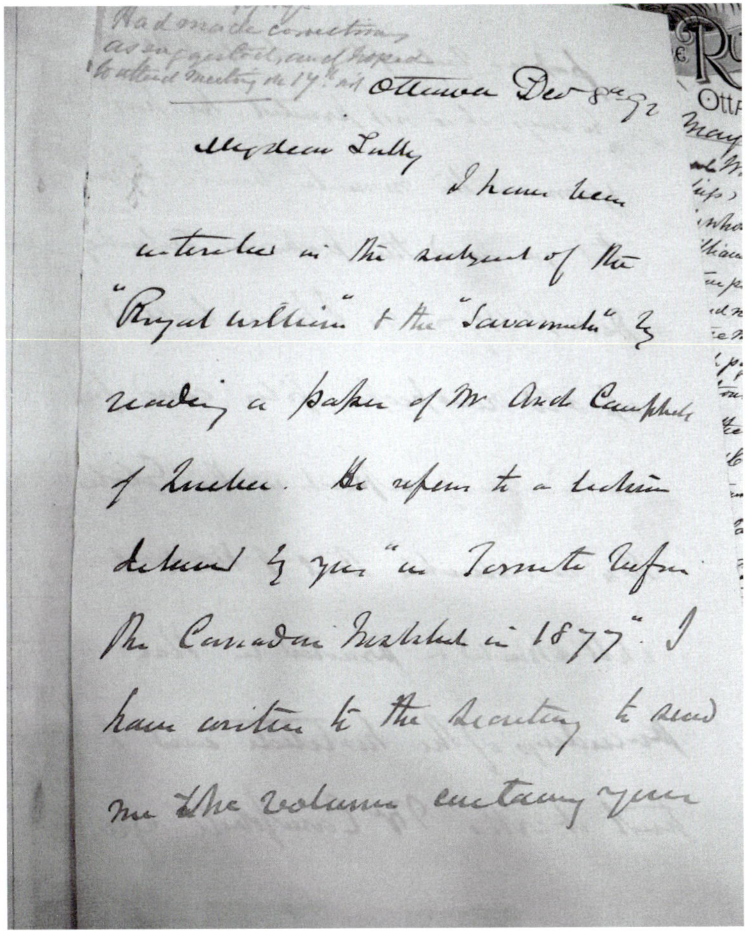
Ottawa Dec 8th 92

My dear Tully

I have been interested in the subject of the "Royal William" & the "Savannah" by reading a paper of Mr Arch Campbell of Quebec. He refers to a lecture delivered by you in Toronto before the Canadian Institute in 1877". I have written to the Secretary to send me the volume containing your

Ottawa, Dec. 8, 1892. Sandford Fleming to Kivas Tully. "My dear Tully. I have been interested in the subject of the "Royal William" and the "Savannah" by reading a paper of Mr. Arch Campbell of Quebec. He refers to a lecture delivered by you in Toronto before the Canadian Institute in 1877 … …" (Kivas Tully, Special Collection, Toronto Public Library)

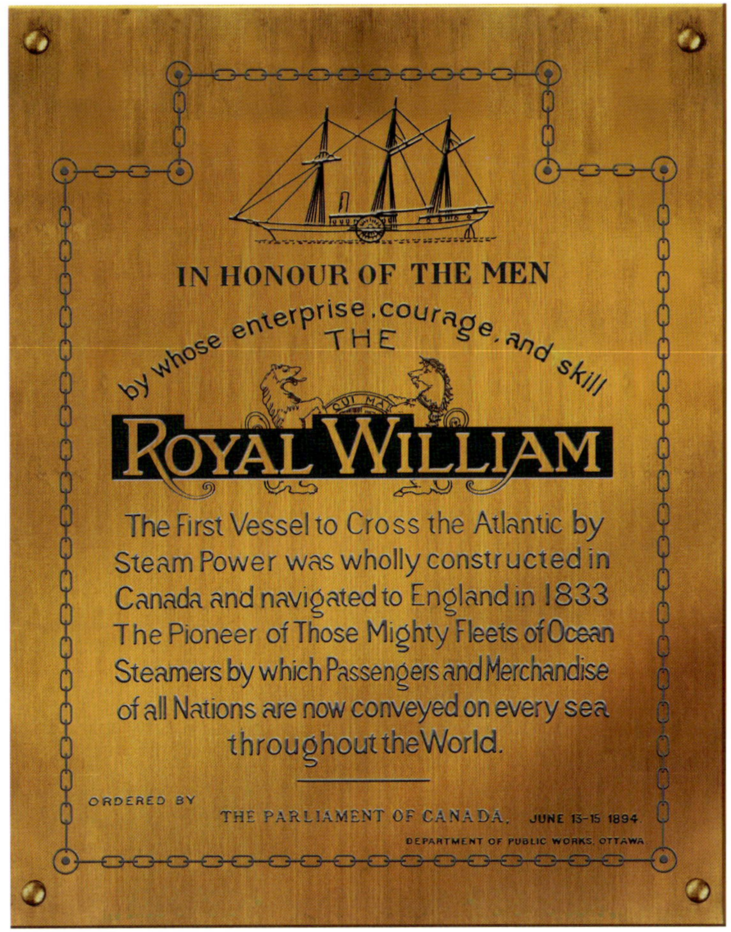

The brass plaque placed in the corridor of the Canadian Parliament Centre Block in 1894 and rescued from the ashes of the fire of 1916. (Reproduction of the plaque, Caroline Soucy)

ocean steamers on which passengers and merchandise of all nations are now conveyed on every sea throughout the world." Ordered by the Parliament of Canada June 13-15, 1894, Dept of Public Works, Ottawa

Thus, Canada emphasized the part the *Royal William* had played in the advancement of regular communication between the colonies and the contribution of steam power to Empire trade, and furthermore to maritime trade throughout the globe.

The unveiling ceremony was attended by many. Those present, besides the delegates to the Conference, included the Earl of Jersey, representing the Queen, members of the Canadian Government, among them Sandford Fleming, the Speakers and Members of both Houses of the Dominion Parliament, officers and members of the Royal Society of Canada, on behalf of the Canadian Institute, the Literary and Historical Society of Quebec, the Nova Scotia Historical Society and associated societies, and citizens of Ottawa generally.

Also present were two brothers, Gustavus W. Wicksteed, Q.C., and Major Horatio Asprey Wicksteed. The former had attended the *Royal William*'s launching and took passage on her trial trip. The latter visited the steamer on her arrival in the river Thames after her epic voyage and was a guest at the banquet given by her captain.

Following the unveiling of the plaque, Gustavus Wicksteed spoke recalling his memories of the vessel's launch.

Horatio Wicksteed
(Courtesy Betty Wickstead)

Gustavus Wicksteed
(Courtesy Betty Wickstead)

No doubt those responsible for the ceremony felt considerable satisfaction, but it took place too late for the key men it honoured. James Goudie, who drew up the final plan for the steamer and oversaw her construction, lived to see the reawakening of interest in her accomplishment in Canada but passed away in 1892. The captain, John McDougall, taught navigation in England for ten years after leaving the *Royal William* and returned to Canada in 1847, hoping to help set up a school of navigation in Quebec, but did not succeed in doing so. Plagued by ill health, he died in 1854.

However, their part in the *Royal William*'s story would live on in the evidence that Frederick Wurtele gathered,

Kivas Tully (1820-1905)
(Victoria Hall. Coburg History, Coburg Internet)

James Goudie (1809-1892)
(Courtesy Ainslie Goudie)

Archibald Campbell (1823-1906) (BANQ)

Sir Sandford Fleming (1827-1915)
(James Craig Annan, artist. Toronto Reference Library)

Frederick Wurtele (1842-1920)
(*The Quebec Chronicle*, March 1920)

which was included in the *Secretary of State for Canada's Annual Report for 1894*, and published in pamphlet form the following year. His final words state its purpose:

This honour has been disputed and the claim is made in Britain that it be awarded to the steamers "*Sirius*" and "*Great Western*," and in the United States to a vessel named the "*Savannah*"; to refute these claims is the raison d'être of this brochure. The *Royal William* was the first veritable transatlantic ocean steamship!

EPILOGUE

The Controversy

Frederick Wurtele may well have made sure that a brochure with the evidence he gathered was published to help prevent or refute future claims, but he soon found out that that was not enough. Even as he was doing so, and as Sandford Fleming was engaged in getting support for his proposal that a plaque be installed in the Parliament Building, an article appeared in the January 1893 issue of the American *Popular Science Monthly* under the heading "The first Transatlantic Steamer."

Wurtele immediately responded in a letter to the editor:

> SIR: On page 424 of the January number (1893) of The Popular Science Monthly is given a précis of the log of the ship *Savannah*, which is correct; but the heading, "The First Transatlantic Steamer," is totally wrong. The *Savannah* was a fraud, a veritable sailing ship, built as such, subsequently took on an engine and propelling contrivances which could only be used in smooth water; with these she steamed out of port, then sailed to England, steaming only eighty hours, not consecutively, out of a passage of twenty-nine days and a half, but took good care to let down the paddles on coming into port, making believe that she steamed.

Far from contrite, the editor wrote in the next issue:

> All that Mr Wurtele says of the defects of the *Savannah* appears on the face of the article we published. The title, if not strictly accurate, reflects current speech on the subject. We are glad to give our Canadian neighbours the credit that is their due in the matter of the *Royal William*-ED.

"Current speech," prevailed in many places as late as in 1944, when a US postage stamp was issued with an image of the *Savannah* entitled, "First Steamship to Cross the Atlantic." Though the *Savannah* helped keep the controversy surrounding the *Royal William* alive for a number of years, now that the truth is known, it is generally conceded that the *Savannah* was not a steamer but a ship with an auxiliary engine that steamed a small part of the way.

As for the other two vessels mentioned in Wurtele's report, these are the facts:

US Postage Stamps of the *Savannah* of 1818 issued in 1944, with the claim that it was the "First Steamship to Cross the Atlantic."

Sirius was a side-wheel wooden-hulled steamship built for the London-Cork route. In 1838, under charter to the British and American Steam Navigation Company, she took the place of that company's steamer *Victoria,* whose construction was behind schedule, and raced the Great Western Steamship Company's paddle steamer *Great Western* across the Atlantic. *Sirius* steamed from Cork to New York in 18 days, and in doing so won the race and, became the first passenger steamer to steam the entire way across the Atlantic from east to west, snatching the honour from the *Great Western* that was purpose-built for the crossing by Isambard Brunel.

Great Western left Bristol four days after the *Sirius* left Cork, due to a fire that broke out in her engine room. In the confusion, not only did Brunel fall and break a leg, which resulted in his having to be put ashore, but fifty of her passengers cancelled their bookings, further delaying the steamer's departure. Nevertheless, the *Great Western* made the crossing faster than the *Sirius*, arriving in New York only hours after her.

But as both their crossings occurred in 1838, five years after the *Royal William*'s crossing, neither of their passages has any bearing on the *Royal William*'s claim.

However, not everyone thought so.

An article entitled "The first transatlantic steamer" in *The Engineer* of August 4, 1893, disposed of the *Savannah* and the *Curaçao* as contenders before describing the career of the

Royal William, and adding, "The *Royal William* can hardly claim to be 'the first transatlantic steamer.' In the light of having been specially built for, and of having continued the service after her first voyage. The first transatlantic steamers in the service were the *Great Western* and the *Sirius,* both of which vessels arrived at New York from the old World on Monday, April 23rd, 1838, and returned again within a week of each other. The *Sirius* did not again cross the Atlantic, but the *Great Western* continued to cross and re-cross, her subsequent crossings being greater triumphs than the first."

It was time that Canada responded, and her answer was the hanging of the plaque commemorating the crossing of the Atlantic by the *Royal William* in 1894.

Then, in 1896, Henry Fry, the veteran Canadian shipowner and Lloyd's agent joined in the Canadian steamboat's defence. In his treatise, *The History of North Atlantic Steam Navigation,* he wrote, "In 1833, Quebec sent forth the first Atlantic steamship, the *Royal William*, which steamed from Quebec to London [via Pictou] in 25 days, four [*sic,* should be five] years before any other ship succeeded in doing it."

He was backed up the following year by Joseph Colmer, Secretary to the Canada Office in London, who read a paper on the state of Canada at a meeting of the Society of Arts in London entitled "The Progress of Canada during the Sixty Years of her Majesty's reign." "Canada," he declared "enjoyed the proud distinction of having built the first steamer that crossed the Atlantic, the *Royal William,* launched at Quebec in 1831."

In 1838, *Sirius* steamed from Cork to New York in 18 days. (Alamy Stock Photo)

He succeeded in convincing some, such as Arthur Pearson, the editor of *Pearson's Weekly,* who wrote under the heading, "The Mistakes We Make":

Great Western, purpose-built to cross the Atlantic, left four days after the *Sirius*. (Alamy Stock Photo)

VI, Some Opinions More or Less Enormous

The First Transatlantic Steamer.

SOME cyclopaedias say that the first vessel to cross the Atlantic by steam was the *Rising Sun* in 1818; others say the first steam voyage was made across the Atlantic by the Savannah. All are wrong. A tablet has lately been erected in the Great Hall of the Parliament Buildings, Ottawa, commemorating the fact that the first vessel which crossed the Atlantic, propelled entirely by steam, was the *Royal William*, built in Canada in 1833 by James Goudie. Some fourteen years previously the *Savannah* crossed from the port of that name to London, but owing to the wood that she carried for fuel running short, she was compelled to cover the grater part of the distance with the aid of sails. And as to the *Rising Sun*, her claim has yet to be proved.

Pearson's Weekly
WEEK ENDING Oct. 9. 1897.

Their words were hotly challenged by an Englishman named James Walker, in a booklet entitled *The First Trans-Atlantic Steamer: An investigation of the claims put forward on behalf of the Canadian steamer* "Royal William" (with notes by Sir C. Purcell Taylor), which was published in London, in 1898. Walker's sarcasm and many inaccuracies suggest that Wurtele's report, followed by the placing of the commemorative plaque in Ottawa, the Canadian Secretary of State's report, Fry's words, Colmer's speech, Arthur Pearson's opinion, and other affirmations of the *Royal William*'s accomplishment, had touched an extremely raw nerve.

As to the inaccuracies in Walker's booklet, he begins by claiming that the first occasion on which the *Royal William*'s claim was made publicly in England was at the meeting of the Society of Arts in 1897. He conveniently forgets that C. Purcell Taylor was the Secretary of the Royal Naval Exhibition Committee, of 1891 that, having examined the *Royal William's* credentials, accepted the model as that of the first vessel to cross the Atlantic under steam, and awarded it a diploma.

He goes on to contend that the *Royal William* was not built expressly as an ocean steamship as "her owners had no power whatever to build or run a steamer for transatlantic voyages, and had they done so, the company would that moment cease to exist in law."

In fact, there is ample proof that the *Royal William* was built with ocean-going qualities. For instance, in a letter to James W. Stewart dated December 1877, James Goudie, the naval architect who designed her wrote:

> In the winter of 1830, Messrs Simmons [*sic*], shipbuilder of Greenock, with whom I served my time, having had applications for the plans, etc, of a steamship to navigate between Quebec and Halifax in summer, and Halifax and the West India Islands in wintertime (I might remark right here that to none more qualified could they have applied, as he had already built four for the Irish trade) … Jas Goudie.

Newspapers, too, such as the *Aberdeen Journal* of 14 December 1831, two years before her transatlantic crossing, referred to the *Royal William* as an ocean steamer:

> It was in contemplation to dispatch the *Royal William* steam-packet to England, by way of showing the possibility of establishing a regular communication with this country by steam, even during the roughest season of the year.

The Quebec and Halifax Steam Navigation Company, the owners of the *Royal William* at that time, abandoned the idea of a transatlantic crossing when it became clear that it would mean that her subsidy would have to be reimbursed if that were done, not because the steamer was unable to succeed.

However, that company had ceased to exist several months before her Atlantic crossing and her new owners had no right to the subsidy, so Walker's contention was neither relevant nor valid.

His facts are no more accurate when he refers to the *Royal William*'s engines. He writes: "The question of greatest interest is the design of her engines, for, within limits, it is on these alone that the success or otherwise of a steamer depend." He believes, he says, that they were built in Canada, and,

> [I]f they were built in Canada, they were built as facsimiles of Boulton & Watt engines that were at the time excelled by other engine builders, such as Maudsley of London, Fawcett of Liverpool and Robert Napier of Glasgow. We may safely conclude that there was not the slightest difference between the engines supplied to Canada (from which these now in debate were copied) and those supplied elsewhere by Messrs Boulton & Watt. I am told by Sir Purcell Taylor, that they were a very conservative firm, introducing very few changes indeed in their designs.

Then, after further denigrating Boulton & Watt engines, he adds:

> It is remarkable that, notwithstanding the number of adulatory reports of her building and launching, no accounts have been preserved of the way in which either ship or engines behaved in Canadian waters. (p.31)

Again, he is wrong on both counts. Copies of the notarized specifications and tender for the engines have been preserved in the archives (See Appendix B), and a photograph of the plans of her engines is in the collection of the Dossin Maritime Museum, and is reproduced on page 38.

Moreover, her Captain, John Jones, RN's account of her behaviour in a particularly bad storm tells us:

> The Steamer was laid to under her mizzen; her paddles moving slowly to prevent lee way and she thus rode out the hurricane with open hatches and without shipping a single sea. Moreover, he added that he, the crew and passengers were all of the opinion that the *Royal William* was to be preferred in a gale to either King's ship or merchantman. (*The Acadian Recorder*, October 8, 1831)

In declaring that the *Royal William* did not travel all the way across the Atlantic under steam alone "without stopping," far from presenting new information, Walker is merely repeating what was well known. The official claim clearly states, as do the majority of other accounts of her journey, that she *laid to* (remained stationary) every few days to have her boilers de-scaled. The necessity of laying to so as to de-scale the boilers in no way invalidated the accomplishment of being the first to cross the Atlantic entirely under steam.

Furthermore, Walker makes much of there having been no special celebrations or lengthy reports of her journey when the *Royal William* arrived in London. If it was deliberate, as he contends, the same explanation that Walker himself suggests regarding Samuel Cunard might apply, though with a different interpretation of Cunard's motive. Walker maintains that had the *Royal William* been "entitled to the least credit for her transatlantic voyage … Samuel Cunard would have published her performance far and wide as a sample of what could be done," adding, "his silence says more than volumes could do." But another interpretation is that Cunard, a shrewd businessman, might not have wanted eventual competitors for a forthcoming contract for the carriage of transatlantic mail by steamers to gain an advantage from the information and perhaps reduce his own chance of obtaining it, which is also probably the reason that the *Royal William*'s logbook was not available for others to see.

And if the omission was not deliberate, the eminent popular British scientist, Dionysius Ladner had made it clear that an Atlantic crossing propelled solely by steam was impossible. "You might as well talk of making the voyage from New York to the moon," he had said. So, who would believe that an unknown steamer had steamed the entire distance across the Atlantic? Moreover, the crossing of the Atlantic under steam had already been celebrated when the *Savannah* crossed, though as it turned out erroneously?

Or might it be that Captain McDougall was fully occupied concentrating on the sale of the steamer, the purpose of the voyage?

Walker then writes:

> Nor has a single line been published from any of the 40 people on board. Letters must have been written home about the voyage, but not one of them is forthcoming. Nor do I find that any steps were taken to ascertain from any of them what the nature of the voyage was. Yet, I am told that some of them are still alive.

If that were the case, and he was so keen on the truth being known, why did he not look those people up himself or ask others to do so?

In his letter to *The Globe* in 1876, J.G. Danter, 2nd engineer on board the *Royal William*, publicly provided his address and offered to provide any information in his possession concerning the voyage to any person interested.

"... The following were the officers on the first voyage:—Commander, Lieut. Jones, R. N.; 1st officer, James Sutherland; 2nd officer, Wm. Pedie; 3rd officer, Thomas Cleghorn. James Laing, Chief Engineer; 2nd engineer, Robert Barsley; 3rd engineer, J.G. Danter; 4th engineer, Jas. Harris.

Captain McDougall commanded her when she sailed for England. I forget the name of his subordinate officers. The Engineer's staff were the following: —James Douglas, Chief Engineer; J. G. Danter, 2nd engineer; J. Short, 3rd engineer.

Regretting the necessity of occupying so much space, any information in my possession will be promptly furnished to any person interested by addressing,

Yours respectfully,
J.G. Danter,
Queen's Hotel, Welland, Ont.
Welland, March 20, 1876.

Yet another of Walker's flagrant inaccuracies is a supposed report in a London newspaper of the *Royal William*'s arrival on the Thames, which is in fact the report of the arrival in London from Quebec of the 3333 ton timber raftship *Columbus* nine years earlier,

He keeps still another astonishing misstatement for the last. "The Canadian *Royal William*," he declares, "was wrongly credited with the successful first transatlantic voyage of another steamship named *Royal William,* which was built in Liverpool for the American and Colonial Steam Navigation Company."

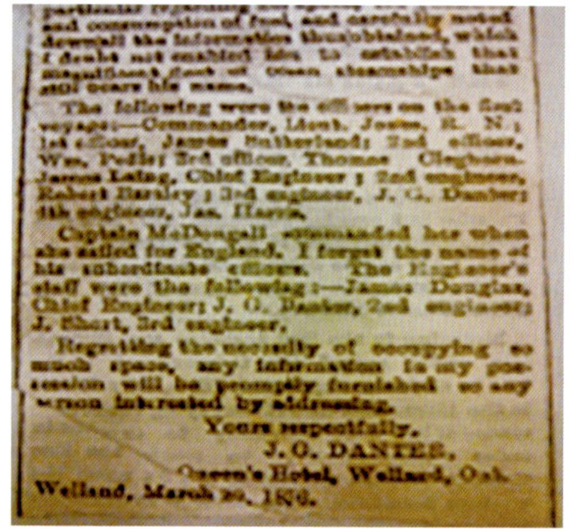

Extract. Letter to the editor of the Toronto *Globe*, dated March 20, 1876, from J.G. Danter, 2nd engineer on the *Royal William*'s Atlantic crossing.

That *Royal William,* under charter to the Transatlantic Steamship Company from the City of Dublin Steam Packet Company, made her first transatlantic crossing in 1838, five years after the Canadian *Royal William*, and even after the crossings of the *Sirius* and the *Great Western.*

Although Walker's arguments were so ill-founded, in the euphoria surrounding Queen Victoria's Golden and Diamond Jubilees in 1887 and 1897, when Britons celebrated so many engineering milestones during her reign, blinded by false patriotism, some insisted that the *Great Western* was the true title holder and that, in spite of the fact that the *Sirius* got there before since the *Great Western* started the race late.

Why was Walker intent on pushing such falsehoods? How many other newspapers carried what the *Huddersfield Chronicle* published on January 15, 1898, and which appeared in the 20th edition of *Chamber's Almanack*?

The Dutch P.S. *Curacao. M. Stoomschif Curaçao*, Willem Hoogkamer, C. Van der Hart.

> Considering the extent to which the Atlantic traffic has grown at the present day the story of its beginning is one of fascinating interest. That it could ever have reached a tithe of its present proportions without steam power is quite certain, and therefore the commencement of trans-Atlantic steaming may be looked upon as the true birth of that great traffic now so regularly and efficiently carried on. The Canadians have claimed to be the pioneers in steam Atlantic traffic, and Mr. James Walker, late of Huddersfield, and now of Ilkley, has in a pamphlet, published by Messrs. Swan, Sonnenschein, and Co. (Limited), of London, subjected this claim to very close and careful scrutiny. Mr. Walker appears to have consulted every possible authority upon the subject, to have scrutinised official reports and out of the way publications, and to have enlisted in the search Captain Sir Alfred Jephson, R.N., Sir C. Purcell Taylor, Bart. (who supplies notes to the pamphlet), and others, and the result of his patient investigation is that this claim is not well founded. Instead of the Canadian *Royal William* being the first to cross the Atlantic under steam the place of honour is given to the *Great Western*, which made the journey in 1838, and was the pioneer of the present day splendid Atlantic fleet of passenger steamers. Mr. Walker give plenty of evidence for the faith that is in him and most readers will hold that he has proved his case up to the hilt.

Did anyone remember on reading articles in periodicals, such as these, that Alfred Jephson was knighted for his services as Secretary of the Committee of the Royal Naval Exhibition of 1891? That committee showed only seven years earlier the *Royal William*'s half-model as that of the "first steamship to cross the Atlantic Ocean."

Not only did periodicals spread Walker's fabrications then, but as recently as 2010, a second edition of Walker's pamphlet appeared on the market to spread them once again. It is due to this new publication that I have thought it necessary to expose the details of his seemingly unending false words.

Here, in Canada, the plaque celebrating the *Royal William* in the Canadian Parliament was joined by two other plaques, one honouring the men who fought in the War of 1812 and the other, a tribute to the Arctic explorer Joseph-Elzéar Bernier. They hung there until the night of February 3, 1916, when the centre block of the Parliament Buildings burned down. Fortunately, the human chain that sprung into action rescued the plaques, which were then stored temporarily at the Parliamentary library and eventually were transferred to the Canadian Archives.

In 1925, the controversy lit up again with a new contender, when the Essex Institute of Salem, Massachusetts published an article written by the American historian Francis B.C. Bradlee entitled, "The First Steamer to Cross the Atlantic. The record of the Royal Netherlands Navy Steamship *Curacao.*"

Bradlee, the owner of an aquatint of the *Curacao,* claimed that she had steamed from Holland to Curaçao and back, in 1827, 1828, and 1829, and that she, and not the *Royal William* was therefore the first steamer to cross the Atlantic.

He dismissed the *Royal William*'s claim because, he wrote, "four years prior to her Western Ocean trip, the *Curaçao* had accomplished, not one sporadic attempt, but three long, regular voyages propelled entirely by her engines, *or as much so as were any of the early ocean steamers.*" (our emphasis)

In those final words, he referred to claimants to the title such as HMS *Rhadamanthus,* which in 1833 got no further than the Bay of Biscay before switching to sail and putting into Madeira to re-coal; the French colonial paddle aviso *Caroline,* ex. *Galibi*, in 1837, whose 6-foot depth did not allow her to carry sufficient coal; the 124 ton steamboat *Cape Breton,* which arrived in Pictou from England after a voyage in 1833 that took 44 days, mostly under sail; and many others, including, as Bradlee himself admitted, the *Curaçao.*

Even if Bradlee had not said so, the evidence is abundant in the reports sent to the Ministry of the Marine in Holland by Captain J.W. Moll of the *Curaçao.* It can be found in the Log Book of her second voyage (see Appendix J), and J.W. Van Nouhuys' *De Eerste Nederlandsche Transatlantische Stoomvart in 1827 Van Zr Ms Stoompakket* Curaçao, *Eerste deel het journal,*

Furthermore, a copy of the entries in the Log Book of her second voyage in 1828 is included in the British *Report from the Select committee on Steam navigation to India, 1834* (see Appendix J). They show that the *Curaçao*, having skirted Europe and Africa, put in to Madeira, before crossing the South Atlantic to Paramaribo, where she took in firewood, and continued to Curaçao. Below it is written:

> The *Curaçao* Packet, commanded by Lieut. Moll, is a vessel purchased in England by the Dutch government and employed between Holland and the West Indies; she is 400 tons burthen and has two 50-h.p. engines. She performed her first voyage from Hellevoetsluis to Surinam in 27 days 11 hours, and Surinam to Curaçao in 5 days, of this she steamed 11 days 3 hours. She returned from Curaçao to Hellevoetsluis in 29 and a half days, steaming on that voyage 23 and a half days. The hour. We have been favoured with an extract of her Log Book for her Second Voyage and, which was performed in 25 average consumption of coal on this voyage was 8 bushels per hour or 6 lbs per horsepower per days and 5 hours, (and shows that her engines were shut down for 10 hours). She returned last autumn from her last voyage commanded by Lieut. LeJeune, who states this, like the two former voyages to be most satisfactory, having encountered several severe gales and hindrances, during which the machinery worked without interruptions or accident. The times of her performing that voyage is not stated, nor her consumption of coal. She was then preparing for a fourth voyage.

Lejeune, her captain on her last return voyage, states that "like her two former voyages," the *Curaçao*'s third voyage was also "most satisfactory," or as Bradlee wrote she was "propelled entirely by her engines, or as much so as were any of the early ocean steamers." Had she been propelled *entirely* by her engines on at least one of her voyages, he would surely have proudly said so. Bradlee's attempt to usurp the *Royal William*'s title is hard to understand.

Conclusion

These then were the *Royal William*'s principal contenders. As in the case of the *Royal William*, none of the other early steamers benefitted from the use of a surface condenser, the invention patented by Samuel Hall in 1834, which desalinated the water that fed steamer engines and prevented encrustations from forming in their boilers.

It was under that shared disadvantage that the *Royal William* succeeded where the others failed. She was able to do so, because she was loaded with sufficient fuel, and because that fuel was of special quality. Commanded by Captain John McDougall, she left Pictou on August 19, 1833, her bunkers and part of her cargo space heavily laden with 254 caldrons of what was regarded in the trade as 'unequalled for steamboats' coal.

And because John MacDougall was an excellent navigator and dared to take on the challenge, when the PS *Royal William* reached England on September 9, 1833, she became the first vessel to span the Atlantic under steam, the First True Transatlantic Steamer.

Appendices

APPENDIX A

Contract and Specification. George Black with the Quebec and Halifax Steam Navigation Company, July 1 and 6, 1830

(Greffe of the notary Lauchlan Thomas McPherson (CN301), ANQQ)

Specification of the proposed steam vessel.
July 1st, 1830 (transcribed as written).

Keel of Rock Elm To be moulded 14 Ins sided 14 Ins in four pieces the scarfs to be 6 feet long each doweled to the other and bolted with 1 In Copper; to have a false Keel of 4 Inches thick fastened also with Copper.

Keelson of Oak To be moulded 17 Ins sided 16 Ins in four pieces bolted thro: every second floor with 1¼ In Copper drove thro: and clinched on a Copper Ring under the main Keel, The scarfs to reach over three timbers, fayed hook and butt tips to be bolted with two Bolts of 1 In Copper, Four Keelsons will be described by the Engineer, with their Iron fastenings, and position as well as the plan of the Boiler hatch, Bulkheads etc and beams over the Engine room.

Transom and Fashion Pieces of Oak Wing Transom sided 16 Ins molded 14 Ins to have two bolts thro the stern Post, with the other Transoms to be in proportion and to have Transom under the Cabin lights secured with Iron Knees sufficiently long and bolted, Fashion pieces to be double sided 9 Inches, the foremast one to run up three feet above the wing transom tip the other, & the other to the under side of the Wing Transom.

Stempost of Oak In one length, with sufficient false post to the same.

Dead wood and Stern Knee Deadwood may be Rock Elm and sufficient depth to give security to the long and half Timber heels, which are to be bolted to the Deadwood to have a knee on the upper piece of Deadwood to scarf under the Keelson & bolted every 20 Ins distance with 1⅛ In Copper.

Stern of Oak To be in Two pieces sided and molded 13 Inches Scarfs to be 3 feet long with well bolted thro: with one Inch Copper and to have sufficient false stern and gripe.

Apron of Oak To be sided 17 Inches moulded 9 ½ Inches and to increase as much as the timbers & to have proper shift.

Sternson of Oak To square 10 Ins & to have proper shift.

Room and Spare To be two feet four Inches.

Floor timbers Oak or Elm Floors to be doubled from 10 feet before paddle Box to 20 feet abaft, and shifted alternately to make the second futtock & laid perfectly solid and caulked 14 Inches square, those forward and aft this space to diminish in size till 10 Inches and to be single floors with a long chock and the heel scarfed on the first futtocks.

Futtocks First or lower Futtocks to be moulded 12 Inches at floor heads in midships, forward and aft diminishing to 9 inches; Midship ones to be Tamarack the others of Oak and sided 14 Inches. Second or middle futtocks to be sided 4 Inches and to be the same description of Timber as the first; Third futtocks to be reduced to 6 Inches thickness at the top 7 and the frame gradually diminishing from the floors heads where it is required to be 12 Inches, third futtocks Red pine and oak. Top timbers red pine.

Timber Moulded At the floor heads 12 Inches, at the wales 9 Inches and at the Gunwales 6 Inches.

Ship to be all Frames & Timber Sided on both sides The Heads and heels of all the Timbers to be properly fayed.

Night heads Oak To be sided 12 Inches & of sufficient length.

Hawse Timbers To have two on each side & to be sided 17 Inches.

Wales & Thick Stuff below To have 5 Strakes of 6 Inches thick by 9 Inches broad. & 2 Strakes of 5 Inches under the wales and 4 strakes of 5 In: at the floor heads all of Oak excepting the floor head strakes (Elm).

Bottom All the bottom to be of 4 Inches of Rock Elm to be wrought in long lengths sound and free from shakes: above light water mark of Oak and 5½ Inches thick.

Shift of plank To be not less than 6 feet and three strakes between butts.

Plank above the Wales To have one strake above 5 In: by 10 In broad and one strake above that 4 Inches of Oak and three Strakes above that of 3 Inch Red pine.

Sheer Strakes To be two Strakes of 4 Inches by 9 In: broad.

Butt Bolts Inside Plank To have two strakes of 4 in next the timber fitted. To have two strakes of 4 in next the outside Keelson bolted thro: and clinched with 1 Inch Copper bolts at least twenty in number on each side with 3 Inch planks above and below the 4 Inch planks the rest of the Ceiling 3 Inches & all of oak above the 4 In strakes, and below of Rock Elm.

Breast hooks Of Oak To be 6 in number 14 feet long sided 4 Inches bolted with 9 bolts in each of 1 Inch in diameter the deck hook to have 11 Bolts and to have an additional one under the bowsprit for security of the Main Stay & Bowsprit.

Slips To be fitted securely in Hog pieces.

Capstan To have as sufficient one fixed with all Ironwork complete.

Treenails To be all of oak well seasoned drove from the outside, wedged and caulked (Canadian oak).

Lower Deck The cabin half deck and forecastle decks to be laid with sound 3 Inch deals & sufficient Beams properly kneed.

Upper Deck To be of the best yellow pine, of 4 Inches thick and 6 Inches wide laid in long lengths free from Sap & Copper nailed with an oak plank next the Combings for the stopper & other Bolts & to have two Bolts thro: the Combings and oak plank to fore and main Hatches clenched before the deck is laid & to have a wide oak plank under the bowsprit heel and one Oak plank next the waterway bolted thro: from the outside every five feet.

Anchor lining To have sufficient above and below the wales and Top Sides.

Pumps etc. To find and fix two metal pumps with chamber & gear complete & to find a spare set of Boxes and Spears, & to build a strong pump well.

Boats, etc. To build Four Boats of Oak viz: a skiff 22 ft. long, a Jolly boat 19 ft. long, and two other boats, the dimensions of which to be given hereafter. Two sets of Iron Quarter Davits complete & Stern davits for the Jolly Boat & to fit all the boats with davits etc complete.

Copper Fastened The ship to be wholly copper fastened below the main wales, the bolts in all the butts to be not less than ¾ In in Diameter & when the sizes are not mentioned to be in proportion.

Caulker's Work To Caulk and stuff all over with a sufficient quantity of Oakum fit for sea and to be tight. To pay the seams of paint work with resin, to scrape & paint both outside & inside with two coats of paint & oil, to plane and paint the bottom green & to black the wales.

Chain plates Fore and main Channels to have preventer plates the fore and main chain bolts 1 ¼ Inches preventer backstays & mizzen to be one inch.

Mast partners To be 6 In: thick by 10 In: sided Oak.

Carlings and Ledges to the Upper Deck To have 3 tier of Carlings to square 6 Inches & to be framed with sufficient ledges to be 3 ½ Inches deep on half beams sided 6 Inches moulded 4 ½ Inches.

Hatchways and Scuttles Hatchways, to be as many as required, with Combings, Hatches, & Iron Bars for the same.

Upper deck Clamps & Ceiling Between decks One strake of 3 ½ Inches under the shelf piece the remainder of 3 Inches under that properly hooked and scarfed.

Upper deck Beams and Fastenings Beams of Oak 12 Inches & five feet apart to be laid on a shelf piece fayed to the timbers, to be 5:(6)In thick & 10 In: broad, to be well bolted thro: every other timber with ⅞ In: bolts & each beam to be bolted thro: the shelf piece & an Iron triangle let into the beam on the under side, to have two wood lodging knees to each end, to be well bolted with the usual number of bolts of 1 In: & placed as required.

Paddle Beams 18 Inches & fastened with 30 iron knees.

Windlass etc. To be of sound Oak 20 In: Diameter properly cased and fitted, and all iron work complete & to have riding chocks fitted, the windlass to be fitted before the foremast with patent palls & patent windlass (strakes).

Catheads To be 12 Inches square & to have Iron hoops on the same & to have bushed sheaves

Carrick and Other Bitts To be sided 6 In: to be of sufficient breadth to carry the windlass, 7 Inches clear of the pall bitts to run down and fasten to the forecastle deck beams or chock between the Beams, the pall bitts to be 16 by 15 In: to run down and fasten to the forecastle deck beams properly fitted and fayed on a four Inch oak plank the whole to be well secured and bolted with I Inch bolts, to have a Belfry, & sufficient Bitts for the Bowsprit.

Plank Sheer To be 4 Inches Oak plank, in long lengths, and to run fore and aft and to bolt thro: every stanchion.

Rails and Bulwark To have sufficient on Quarter and Main Decks with Compass rail across the Stern Double bulwarks from the fore Channels forrwrd to be of 1 ½ In: Oak plank. Abaft that to be 1 ¼ In: rabetted deal.

Head To have a Head with chocks, rails etc complete.

Rudder Tiller Wheel etc. To have a sufficient rudder and Tiller with 3 pairs composition pintles & braces and one dumb cleat and two pairs of Iron pintles and braces all of sufficient sizes & properly fitted, bolted & spiked & to have two Iron hoops in the Rudder head & to find and fix a Mahogany Steering Wheel, brass mounted.

Anchor Stocks To find Anchor Stocks & to stock the Anchors with sufficient Iron hoops to the same.

Fore main And Mizzen channels To be 5 Inches at the side & 3 ½ Inches at the outer edge, properly bolted and complete & with rails & to find dead Eyes & binding of sufficient size with preventer plates to every shroud and back stay plate.

Hull completed for masting To be fitted complete with all necessary Davits cleats, Rails, pins & sheaving attached thereunto ready for masting.

Fastenings to Gripes & after Part of Keel To have ring plates to the Gripes & dovetailed plates to the after part of the Keel all of sufficient size and composition metal.

Iron Work To find and drive eye bolts for the standing and running Rigging, Stoppered Gun bolts & Shank painters. The bolts in all the Knees, breasthooks etc to be well clinched on rings & the stopper bolts to be 1 ¼ Inch diameter.

Plumber and Glaziers work To find and bore 8 scuppers and two hawse pipes to lead under the Catheads and Bowsprit and in the way of the night heads to lead the seams under the knees of the head & hawse Boxes and to Glaze the stern windows, Skylight & Binnacle.

Joiner's work To Plane the Bottom, sides & wales, underside of Upper deck Beams & Shelf pieces.- to emboss the ship outside & inside all fore and aft as customary to build a companion, a Sky-light & Stern Windows, to build a strong Cambouse house on deck & a stand for the fire hearth the lower counters to be 3 in oak plank & the berthing of the stern to be 2 in Deal rabetted with deadlights to the same carved Quarter pieces quarter Galleries & cove according to the plan. The bottom after being properly planed to be painted Green with a sufficient number of Coats.

The builder to find a complete set of yards, Masts, Spars, tops, Cross trees Steering sail booms etc etc Complete. - The Lower Masts and Bowsprit to be of Yellow pine. The Top Masts Lower Yards & Jib boom to be of Red pine & all the other Masts, yards, Spars etc to be spruce & to be all made of the proper dimensions. To find a set of ash oars Rudder & Tiller for each Boat & the Boats to be painted inside & outside with paint & oil, To find a set of Hickory handspikes & Capstan bars, the whole of the masts, spars, yards etc to be finished or to have a coat of bright varnish. To make & fit a Fife rail & stanchions round the main mast & pumps & a rail across the Bowsprit inside the & fit the rail stanchions etc fore & aft with belaying pins complete.

The ceiling To be caulked to within 8 feet of the keel.

From the Break on the foreside of the paddle Box, the Ceiling and outside planking are to be carried forward to the Beam before the paddle Beam up to the Deck & diagonal Timbers from the beam to the frame passing between the two skins & bolted through and through.

The Stern To be 14 In. at rabbet 3 In: outside.

A stringer or thick plank six feet from Gunwale 14 Inch by 8 Inches bolted to every frame & at every butt up as far as the Bends one bolt is to be put through each plank end & clinched on the Ceiling. Floors are all to be bolted through with 1 ¼ Inch Copper.

The cutting down line to be two feet instead of 18 Inches as mentioned in the plan.

The vessel to be entirely built in frame & each frame to be bolted to the preceding from the stern to the Bow with ¾ Inch Iron bolts at each 3 ½ feet distance from Keel to Gunwales.

The ship to be delivered afloat tight and complete in all the month of April 1831.

Quebec 1st July 1830

The rudder to be straight and the curve abaft the paddle wheel to be according to the body plan, in other respects the vessel to be built according to the plan exhibiting the water lines. Each tender to specify a certain sum for the hull & constructed according to this specification & to the draught shown (without reference to Tonnage) to be built under the inspection of any person the Committee may appoint. The person whose tender shall be accepted to bind himself to furnish a model according to the draft before commencing the Building.

By Order (signed)

Wm Stevenson, Act Secy

Quebec 6th July 1830

Copper for fastening, Iron Knees, Hawse pipes & Patent Windlass to be imported by the committee & the Contractor is bound to receive the same in payment at cost and charges. Wm. S.

The specification mentioned and referred to in the contract and agreement to which the same is now annexed, as forming part of the said

Contract and Agreement.

(Source: Vancouver Maritime Museum Archives VMM Reference Files collection "Ships" Royal William - Ship Files Original builders' sp) cifications for the *Royal William*, SF2005.001.0659, July 6, 1830)

APPENDIX B

Specifications and Tender for engines: Bennet and Henderson with the Quebec and Halifax Steam Navigation Company, August 28, 1830.

(Greffe of the notary Lauchlan Thomas McPherson (CN301), ANQQ)

Specification of the Engines etc. (transcribed as written) The Engines and Machinery generally shall be made according to the Drawings unless some alterations should be suggested by the persons making tenders and approved by the Committee, It being well understood that those parts of the detail of construction which are not shown in the Drawings shall be made according to the most approved mode adopted in other Boats or according to plans to be furnished by the Committee.

The Cylinders shall be fifty Inches diameter and the length of stroke five feet.

The air pumps shall be lined with brass and the barrels of the hand and connector feed pumps shall be also of brass as shall also all the Valves and Seats of valves throughout the Engine. The feed pipes and the injection pipe, if required shall be copper, and generally all the working parts exposed to the action of the water shall be of brass or copper.

There shall be a continuance to such Engine to cut off the steam at any point of the Stroke from ⅓ to ⅔s.

There shall be Two Safety Valves one of which shall be locked up and inaccessible to the Engineer.

There shall be a Copper pipe of about one Inch in Diameter going down to within 4 Inches of the upper side of the flue and rising 16 feet above the top of the Boiler for the purpose of giving notice of any dangerous deficiency of water in the Boilers.

There shall be proper brass Cocks etc for emptying and cleaning the Boilers and there shall be a continuance for allowing a certain proportion of the feed water to escape, constantly and the feed pipe shall be carried for the distance of 10 feet thro: the pipe by which the water shall be so permitted to escape.

The Cranks, Shafts, Crossheads and Cross Tails shall be of wrought Scrap Iron. Should the Contractor in preference to making in this Country these parts of the Machinery, choose to import them & to order Two Sets to avoid the risk of disappointment and that both sets arrive, the Company agrees to pay the cost of the duplicate set and the changes incurred in bringing it out.

There shall be Two hand feed pumps of sufficient size to supply the Boilers with Water when the Engines are not at work and the Feed when the Engines are not at work shall be regulated by a stone float,

There shall be Two Eccentrics to each Engine and the Engines shall be so made that they may be disconnected by removing the Drag Link and pin. The floors of the Boilers shall be 3/8 In Iron and the shell of 5/16. In both of the best quality which can be had in this Province.

There shall be made a brass barrelled double acting force pump 8 in in diameter and Two feet stroke to be connected with the Engines, at pleasure & & having an air vessel and stand pipe of Copper or brass with proper nozzles & screws for fastening on leathern hoses the said pump acting also as a bilge pump when necessary.

There shall be Two Boilers of 25 feet long each & 12 feet wide according to plan.

The workmanship of the whole Machinery shall be at least equal in every respect to that of the Engines of the John Molson and complete.

The whole to be completed fit for work within Two Months of the date at which the Boat will be delivered to the Contractor at Montreal, under a penalty of Fifty pounds per day for every day taken beyond the time specified.

Tender for engines:

Montreal, 7th August 1830
The Committee of Direction of the Quebec and Halifax Steam Navigation Association
Gentlemen,

We offer to construct the two steam engines required by you according to the specification and plans exhibited in your committee rooms in Quebec on the 1st inst. viz. two engines with cylinders each to be fifty inches in diameter, of five feet stroke – the engines to have side beams with iron framing the air pumps to be lined with brass - brass feed pumps- rods and valves and seats of valves to be also of brass – feed pumps and injection pipes to be copper if required to cut off the steam two safety valves, one whose off to be out of the reach of the engineer, a copper pipe one inch in diameter to reach sixteen feet in height as a warning pipe, proper brass cocks for supplying and cleaning the boilers – two hand feed pumps for supplying the boilers – two eccentrics – a brass barrel acting force pump of eight inches diameter and two feet stroke to be worked by the engine with an air vessel to act as bilge pump.

All the cranks, shafts, crossheads and cross tails to be of wrought iron – the two boilers to be of iron with flues 3/8 an inch thick and shells 5/16 of an inch thick.

These engines to be erected in a vessel to be furnished by the Committee at Montreal next spring within two months after such vessel shall be delivered and the whole machinery will be put by us into perfect good working order all for the price of Six thousand six hundred pounds currency

As the committee is desirous of having it in their power to reject the boilers to enable them to adopt a new principle in contemplation should therefore the rejection of the boilers take place the price of the engines will be reduced to four thousand eight hundred pounds?

We beg to remark that the difference in this plan over former tender under date of the 16th ult. is occasioned by the wrought iron shafts and brass pumps.

We offer as our securities the Hon. Chas Wm Grant and Thomas Busby Esq., both of this place.

Price including boilers	£6,600
" without boilers	£4,800

We have the honour etc Bennet & Henderson

These two documents are attached to the contract between John Henderson & John Bennet and various stockholders of the Quebec and Halifax Steam Navigation Company to make machinery.

Gr. L. MacPherson, 28 August, 1830, ANQQ

Notes to Appendix B

The paddlewheel is not mentioned in either of these specifications, for which there was presumably a separate contract that has not yet come to light.

According to James Goudie:

> Her model was quite different to any ordinary ship or steamer. Her shape of frame was precisely that of a "bell," with the mouth turned up, thus forming a recess for the wheel, at which part she was but 29 feet from outside to outside, while at either end of the paddles she was 17 feet, making her tonnage old measurement, over 1,600 tons, instead of 363 tons as stated.
>
> It was an error in the customs of that day, although I tried to convince Mr. Dunscombe of his error, but could not alter his opinion of measuring between the paddle wheels, which did not extend any further than the diameter of the paddles, which was in a recess formed for the protection of the wheels.
>
> The wheels had double flanges and arms on each of the flanges, 8 feet floats, 23 feet diameter. The shaft was supported by a beam fore and aft, on the ends of the cross beams, forming the opening for the wheel.
>
> The deck extended from one side to the other of the widest part of the ship, except in the space occupied by the wheels. (*Quebec Morning Chronicle*, 24 May 1888)

APPENDIX C

Contract for building the cabins of the steamship between Joseph Andrew with the Quebec and Halifax Steam Navigation Company, January 22, 1831

(Greffe of the notary Lauchlan Thomas McPherson (CN301), ANQQ)

James Goudie to Quebec Morning Chronicle, 24 May, 1888.
Specifications in the contract awarded to Joseph Andrew, carpenter and joiner of Montreal to finish the interior of the *Royal William:* (transcribed as written)

Frames of the Stern Windows and covering of the Counter to be of pine, and to have a brass rail five inches high round the fore part of the Counter covering three inches from the edge.

The back of the Sofa at the Stern to be of Mahogany panelled and to have arms of the same material according to pattern.

All the stairs to be of Oak. The balustrades and rails to be of Mahogany, with one half-pillar (turned) in the centre of each scroll, at the foot of the stairs.

The joiner's work of the two water closets to be done complete with oak, except the seat, which is to be of Pine.

The Pilasters between each of the State Rooms to be fluted, the front part of each to stand out 1½ inches from the face of the framework, and to be six inches broad with a capital according to the drawing herewith shewn.

The Doors of the State Rooms and the space between the Doors and Pilasters, to be panelled according to the drawing. The State Room doors to be sliding hung from above and to have best mortised locks and handles of brass.

The ceilings, sides of Berths, scuttles and beams to be lined with Pine boards grooved and tongued the lining boards to be 6 inches broad, with a small beed between each joint.

Each State Room to have two Berths (with the exception of the two aftermost which have only one each) and one plain wash handstand of Mahogany.

Complete the front of each Berth to be panelled, the bottoms to be of sacking or canvass, and each State Room to have a small bracket-table covered with green cloth, and fixed at the opposite end from the wash handstand.

The State Room doors (in place of the top panel) to have in each a *Jalousie* to open inside with small brass hinges and button, to fasten.

The shifting Partitions to be panelled and made with Hinges or Slides so as to shift from one length of Berths to another at pleasure.

The Middle Berths to be placed according to [plan, and to be of corresponding materials, workmanship and appearance with the rest of the Cabin.

Upper cabin

To have four good substantial panelled doors, with strong hinges, handsome mort iced locks and handles of brass.

The State Room in each side of the Cabin to have two panelled doors, to have a Berth in each, the front of which shall be panelled, and the bottom of canvass, the Doors to have best mortise locks and handles of brass.

Two other doors (as shewn in the plans) with mortise locks, hinges, and handles of brass. These are the pantry doors.

The interior of the upper Cabin to be panelled, from the base to the surboard? And above that to the cornice to be planed and prepared for papering.

The whole of the wood used in fitting the Cabins to be of the very best description of pine, well seasoned, except in cases where another description of wood is named and all the material used in fitting up the Cabins to be of the very best description.

The work to be subject to the inspection of any person the committee may approve for that purpose.

APPENDIX D

Ratification of a contract made in Montreal on March 19 1831, between Charles Fry and William Stevenson, April 23, 1831

(Greffe of the notary Lauchlan Thomas McPherson (CN301), ANQQ)

Charles Fry, cabinet maker and upholsterer of Montreal, was to provide all the cabin furniture and furniture for the berths: (transcribed as written)

30 mattresses made of the best description of hair to fit the several berths in the said vessel beginning from the stern and to be of similar description as those being made for *John Bull*, each mattress to weigh about 25 lbs and completely finished at the rate and price of 2/- currency per pound weight.

Fifty-two pillows of the best live Goose feathers, similar to those making for the said *John Bull*, each pillow to weigh about four pounds, at the rate of 3/9d per pound weight.

Curtains for the State Rooms if of worsted damask to be according to pattern furnished to the said Stephenson with edging complete and fitted up at the rate of 35/- per single Berth, and if of *Moreen* (ribbed like corduroy, but can be made of cotton, wool or wool and cotton) at the rate of 28/6 per single berth optional on the part of the said party of the second part.

Two Mahogany single headed Couches with scroll at one end, each to be two feet wide and six feet long on the cushion, stuffed with hair, and covered with hair cloth, the cushions to be moveable, the Couches to be handsomely carved, and to have a contrivance at each side so as to convert the same into a bed at pleasure (being intended for the ladies' Cabin) at the rate of £11 5s each.

A Hair cloth cushion for the Stern seat to be 9 feet long by 20 inches in breadth at four pounds five shillings. Twelve rush bottom chairs and two arm chairs to match, to be counted 15 chairs, of a handsome pattern, at the rate of 10/- each.

A set of Mahogany dining tables to be 28 feet in length by four feet two inches wide, with sliding frames at each end, to be fitted to the mizzen mast, and to have two shifting flaps at each end, the whole made of sufficient strength for a sea-going vessel, with iron fastenings to the deck for the sum of £45 currency.

Four Mahogany side tables with brackets to be hung to the side of the round house, each 6 feet in length by two feet in breadth, to fold down and to be moveable with all fixings and all clamping complete, for the sum of £12 10s for the four tables.

And four rush bottom sofas or settees of 6 feet long each by two feet wide with backs and arms complete to match the said chairs and to have a contrivance in front so that each may be turned into a bed at pleasure for the price of fifty-five shillings each.

Twelve chairs and two armchairs to match the said four sofas or settees to be of handsome pattern and to be strong and fit for a sea-going vessel at the rate of 10/- for each of the said twelve chairs and fifteen shillings for each arm chair.

Two handsome Mahogany sideboards with flush tops and a back each, to have three drawers in the upper part, to have pedestal closets, the one fitted for wine- bottles, the other with three sliding shelves, to have a cupboard, in the centre with two doors, the pedestals to be with half-columns for the price or sum of £15 each.

The upper Cabin floor to be covered with the best description of floor oil cloth and of an improved pattern, at the rate of seven shillings and six pence Currency per square yard laid down.

The ladies' Cabin floor to be covered with the best Brussels carpeting well made up and laid down at the rate of 6/1½d for each running yard, and all the Berths or state Rooms to be carpeted with the best Scotch carpeting made up and laid down complete, at the rate of 4/6d per yard.

The said Charles Fry shall furnish every other article in the cabinet and upholstery line that may be required for the said steam vessel at and after the same rate and price of articles of similar description new or to be made and furnished by the said Charles Fry for the aforementioned steamboat *John Bull* and in no case shall a higher price be charged.

APPENDIX E

Registration, Quebec, August 22, 1831

Description of the PS *Royal William* according to her registration certificate issued on the 22nd August 1831, two days before she left Quebec for Halifax on her first trip:

Burthen 363 60/94 One Deck + Round House, 3 Masts, length 160 feet, Breadth taken above the Main Wales 28 feet, Depth in Hold 17 feet 9 inches, schooner rigged with a standing Bowsprit, square sterned, carvel built, Quarter Badges, scroll Head, Admeasured afloat. Propelled by steam with wheels or Flyers at each side.

The measurement of her breadth was taken inside the paddle-wheels. In her subsequent registration, the outside measurement was taken, i.e. 44 feet.

Quebec 22d August 1831

Register Steam Ship

Royal William

AB

No. Forty two Port of Quebec dated 22nd August 1831 ~~185~~

Name Royal William Burthen 363 60/94 Tons John Jones Master.

When and where built or condemned as Prize, referring to Builder's Certificate, Judge's Certificate, or Last Registry.... } built at this Port this present year 1831 — which appeared by a certificate of Geo. Black the builder, dated 15th July last.

Name and Employment of Surveying Officer (signed) C. G. Stewart

One Decks, Three Masts, length 160 feet —— inches, Breadth, taken above the Main Wales 44 feet —— inches, Height between Decks, —— feet —— inches or Depth in Hold, 17 feet 9 inches Schooner Rigged with a Standing Bowsprit, Square sterned, Carval built, Quarter Badges ~~Galleries~~, Scroll Head, that the Frame Work, and is of and that she is a

admeasured aground — Propelled by Steam with wheels or Flyers at each side
Breadth between the Paddle Boxes, Twenty-eight feet.

NEW FORM.

Decks, Masts, that her length from the inner part of the Main Stem to the fore part of the Stern aloft is Feet Tenths, her Breadth in Midships is Feet Tenths, her Depth in hold at Midships is Feet Tenths, that she is Rigged with a Bowsprit, Sterned, Built, Galleries, Head, that the Frame Work and is of and that she is a

Subscribing Owners	Shares
William Finlay, William Walker and Jeremiah Leaycraft of Quebec, Merchants, Trustees of the Incorporated "Quebec & Halifax Steam Navigation Company."	Sixty-four

Other Owners

Custom House 14th June 1832

William Finlay, William Walker and Jeremiah Leaycraft of Quebec, Merchants, Trustees of the Quebec and Halifax Steam Navigation Company have assigned by deed of Mortgage dated this day 64. 64th Shares to Sir John Caldwell, Matthew Bell, Jeremiah Leaycraft, Noah Freer, James Bell Forsyth and Henry Lemesurier of same place, Merchants.

Custom House 15th May 1833

Seized by the Sheriff of the District of Montreal and sold by him to Matthew Bell, Noah Freer Henry Lemesurier, Jeremiah Leaycraft, Henry John Caldwell and James Bell Forsyth of Quebec, Merchants, by Bill of Sale dated 4th May Inst.

De Novo Nov. 13. 1833

A true copy.

Custom House Quebec

2 March 1891

S. D. O'Meara

acting Registrar of Shipping

APPENDIX F

The 235 Founding Shareholders of the Quebec and Halifax Steam Navigation Company

Allen, Thomas C.
Allen, William M.
Allison, Joseph
Anston, Joseph
Archibald, Edward M.
Avery, James T.
Aylwin, Charles
Bain, James
Barron, John
Barry, John Alexander
Barton, George
Belcher, Andrew
Bell, Francis
Bell, John
Bell, Matthew
Black, George
Black, James
Black, Rufus
Black, William
Blackstock, Richard
Boggs, J.
Breckenridge, James
Brehm, William
Bridge, James
Budden, William
Caldwell, John
Campbell, Archibald
Campbell, John Saxton
Cariff, William
Carter, William
Clark, Frederick H.
Clarke, Alexander
Clarke, Christopher
Clearihue, James
Cowan, William jr
Cramey, Martin
Creighton, J.G.A.
Creme, Andrew
Cunard, Henry
Cunard, Joseph
Cunard, Samuel
Dalkin, Robert
Daly, Dominic
Darby, Joseph
Davie, Allison
Davie, Richard
Davis, Samuel
Dawson, Robert
Deblois, Edward
Denston, Robert
Dewolf, Charles
Donaldson, James
Douglas, John
Downes, Robert
Dubord, Hyppolite
Duchesneau, Adam
Durette, Francis
Dyke, Joseph
Eade, William
Epson, Adam
Fairbanks, Charles
Finlay, William
Forsyth, John
Foster, William
Fraser, Alexander jr
Fraser, Andrew
Fraser, John
Fraser, John Malcolm
Freer, Augustus
Freer, Noah
Gates, Horatio
Gibb, James
Gibb, Thomas
Gordon, Thomas
Grant, Charles William
Grant, Thomas
Grassie, Thomas
Gugy, Lewis
Hamilton, James
Hardley, George
Hare, David
Hartshorne, George
Harvey, Richard Sr
Harwood, Robert N.
Hawbolt, John
Henderson, Alexander
Henderson, Gilbert
Henderson, Patrick
Henderson, Robert
Henderson, William
Holland, Philip J.
Holt, Charles A.
Howe, John
Hunt, James
Innes, George
James, Edie

Johnson, J.A
Johnson, John
Johnson, John Jr
Jones, John, Jr
Jones, Nathaniel
Keefler, Charles
Keith, Alexander
Keith, Daniel
Kerr, John
Key, Alexander
Keys, George
Lambly, John
Lampson, William
Lawson, George P.
Leather, John
Leaycraft, Jeremiah
LeCain, Francis
LeCain, Nicholas
Leggate, Eliza
Lemesurier, Henry
Letson, James
Lishman, James
Little, Mary
Locke, William
Lockyer, Henry
Logan, David
Logan, Hart
Macara, William
MacDonald, James
Marshall, Richard
Maynard, Thomas
McCallum, Colin
McCallum, Duncan
McCully, Caleb
McDonald, Alexander
McDonald, Allan
McDonald, James
McGill, Peter
McGregor, Alexander
McKenzie, James
McLeod, John
McNab, James
McNab, Peter
McNider, Adam L.
McQuillan, Edward
Miller, John
Mitchell, James
Mitchell, Samuel
Moffatt, George
Mordor, Joachim
Morison, Alexander
Morrell, Hugh
Morrison, Alexander
Mortimer, William
Munn, John
Munro, John
Neilson, Samuel
Patterson, Peter
Patterson, Robert
Peabody, Francis
Pemberton, George
Pemberton, Henry
Pemberton, William
Peters, Thomas H.
Philips, William
Piers, Temple Lewis
Potter, Edward
Price, William
Primrose, Alexander
Primrose, J.
Prior, Henry
Racey, John
Rainnie, John S. Gavin
Randall, George
Rankin, Alexander
Ratchford, John Jr
Richardson, John
Richardson, Robert
Rigby, Charles
Ritchie, James
Ritchie, William
Robb, James
Romans, John
Romans, Robert
Ross, A.P.E.
Ross, Patrick
Ross, Robert Pope
Rouast, Jasper
Roy, Charles Francis
Roy, William Henry
Russell, George
Russell, Hector
Russell, Joseph
Ryan, John
Samuel, Joseph
Saunders, James
Schneller, Benjamin
Shannon, J.
Shaw, Andrew
Shaw, Joseph S.
Shaw, Robert
Sheppard, William
Shirreff, Alexander
Shortis, Robert
Simpson, Alexander
Smith, George
Stairs, James L.
Stairs, William
Stanfield, James
Starr, Danie
Starr, Daniel
Starr, David
Starr, Joseph
Stayner, John
Street, James A.
Taylor, George
Taylor, George
Tidmarsh, James H.
Tobin, J.
Tobin, Michael
Torrance, Benjamin
Torrance, John
Tremain, Jonathan
Tremain, Richard
Tucker, Thomas
Turner, George
Urquhart, Margaret
Walker, William
West, Conrad
Wilkie, James
Willard, Asa
Willaston, John S.
Woodill, William
Yates, L.
Young, George
Young, William
Young, William F.

APPENDIX G

Extract from a report of the committee of the Quebec and Halifax Steam Navigation Company (transcribed as written)

The Sydney Herald, New South Wales, 29 November, 1832

The Committee, with an earnest desire the vessel should, in every respect he suited to the proposed navigation, "procured plans from various parts of Great Britain at an expense of seventy guineas, £80. Upon the most approved of these, the Hull of the steam vessel was contracted for to be ready to launch in the course of the month of May, for £5475 currency, and there has been various extra work performed by the builder, that will, in all probability, amount to £150 more, and making for the hull, boats and spars, £5625.

The Committee have also contracted for the following work, viz. -For the two engines, of about 90 horse power each, of the best materials, £6,600.

For the joiners work of the cabin, £510.

For Painters' and Glaziers' work, £42 1s.

They have purchased the chain cables & anchors necessary, amounting to £373 19s. 5d.

They have also made arrangements for the supply of sails, rigging and fitting the same, probable cost, £600.

For the upholsterers work, furniture of the cabins bed and table furniture, will, in all probability, amount to £900.

The Committee has agreed to pay an engineer for drawing plans, superintending the construction of the engines, and seeing the same fitted in the vessel, £150.

For superintending the building of the Hull, £60.

For the services of an assistant to the Secretary, £12°- Jfc.?

For stationery, books, advertisements, postages, &c. &c, £39 13s. 2d.

In addition to the above list of known expenses, there may be others that cannot as yet be estimated by the Committee, and which may probably amount

Total amount, £15,607 2s. 7d.

The Committee further report, that they have engaged a commander for the vessel, at a salary of £250, for the first year, to commence from the date of launching.

That they have appointed agents for Quebec, Halifax, and Miramichi, from each of whom security is required, the agent at Quebec in the sum of £1,000, and the others in the sum of £500 each. The accounts of the different agents to be rendered to the Treasurer and Secretary monthly.

Under the impression that it was their duty to secure the shareholders from all risks until they should have an opportunity of insuring their separate interests, the Committee have ordered insurance to be effected on the Hull of the vessel valued at £6,000, currency, from the time of launching until the first of July, the period specified in the different contracts for the vessel being completed, but as it is not the intention of the Committee to insure any sum beyond that period, It will be for the different stockholders to insure their separate interests, should they see fit.

The Committee beg to lay before the meeting a copy of the Act of incorporation, under which the Shareholders are called together, for the purpose of electing a Committee, for the management of the affairs of the Company, and also Trustees, in whose names the vessels shall be registered.

APPENDIX H

Registration of Steam Ship *Royal William*, Quebec, 22 May 1833

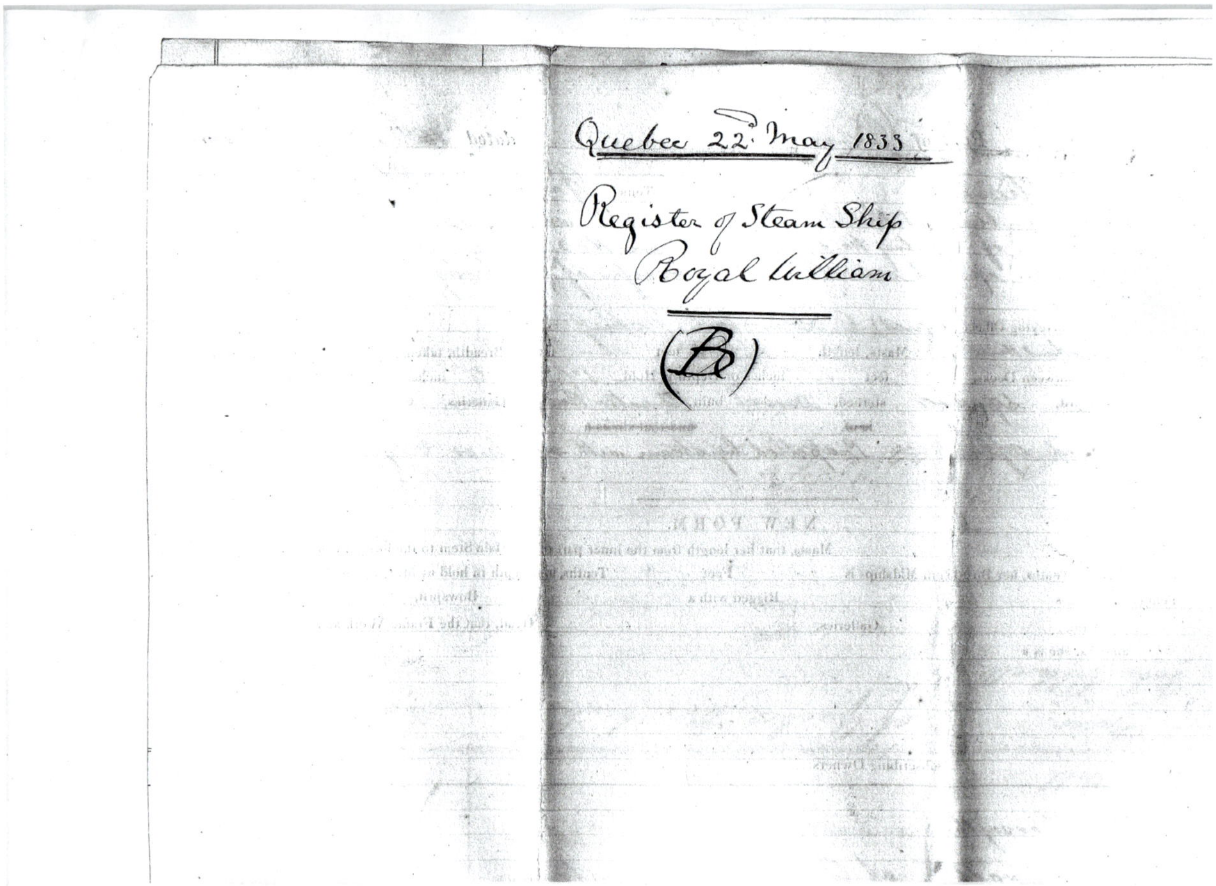

Quebec 22d May 1833

Register of Steam Ship
Royal William

(B)

Copy

No. *Thirteen* Port of *Quebec* dated *18th May 1833* ~~183~~

Name *Royal William* Burthen *363 60/94* Tons *John McDougal* Master.

When and where built or condemned as Prize, referring to Builder's Certificate, Judge's Certificate, or Last Registry.... } *built at this Port in the year 1831 – which appeared by a former Cert: of Registry No. 42. granted here the 22nd Augt 1831 – now delivered up and cancelled upon transfer of property*

Name and Employment of Surveying Officer *(signed) C. Secretan, acting*

One Decks *& Round House* *3* Masts, length *160* feet — inches, Breadth, taken *above* the Main Wales *44* feet — inches, Height between Decks, — feet — inches or Depth in Hold, *17* feet *9* inches *Schooner* Rigged with a *standing* Bowsprit, *square* sterned, *carval* built, *Quarter Badges* ~~Galleries~~, *scroll* Head, ~~that the Frame Work, and~~ ~~is of~~ ~~and that she is a~~

Admeasured afloat – Propelled by steam with wheels or Paddles at each side

NEW FORM.

Decks, Masts, that her length from the inner part of the Main Stem to the fore part of the Stern aloft is Feet Tenths, her Breadth in Midships is Feet Tenths, her Depth in hold at Midships is Feet Tenths, that she is Rigged with a Bowsprit, Sterned, Built, Galleries, Head, that the Frame Work and is of and that she is a

Subscribing Owners	Shares
James Bell Forsyth } *of Quebec, Merchants*	*Ten*
Jeremiah Leaycraft }	*Ten*

Henry Lemesurier		Ten
Other Owners		
Matthew Bell Noah Freer Henry John Caldwell	of Quebec, Merchants	Fourteen Ten Ten

De novo London 22nd Novr 1833

A true copy
Custom House Quebec
2 March 1891
D. D. O'Meara
acting Registrar of Shipping

Pictou Custom House register, MG 100 vol. 227 no. 8d, Nova Scotia Archives, Halifax.

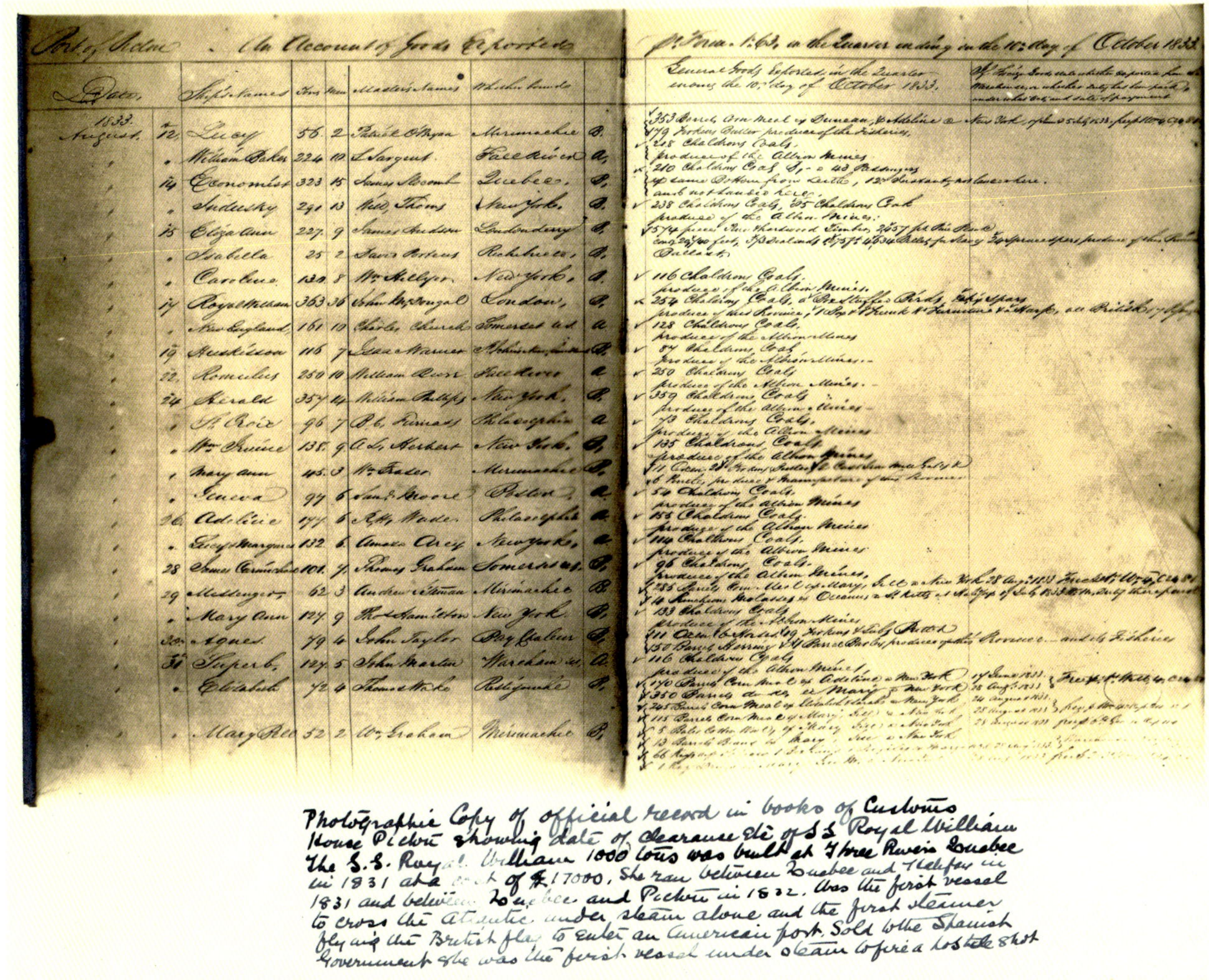

APPENDIX J

Observations on the weather and other occurrences on the PS *Curaçao*'s 26 day passage in 1828 from Hellevoetsluis, Holland, to Paramaribo, Surinam*

- - Fresh breeze with showers; started at ½ past noon; at 3 outside the Narrows; at 4 o'clock S.E. by E; 61 men on board, passengers included. Draught of water, l3 f. 7 inc. and 12 ft. 5 inc; wheel diameter 15ft.; on board 360,000 lbs. coals.

1. Engines working very smooth.
2. Strong wind, showers, bad weather; at 4 in the morning S Foreland in sight; at 7 ½ unshipped the Pilot at Dover.
3. Light breeze, good weather; at day break of Start Point; great produce of steam, water good for drinking.
4. Breeze freshening to tight- reefed top sails; sea very rough; much rolling in the night; sight of the Ouestpant light.
5. Top gallant breeze; difficult swell of sea; a brig in sight, upon which gained very little.
6. Light breeze; fine weather; discovered a small leak in the boiler, S. E.; wheel 17 feet diameter.
7. Fresh breeze, fore top-mast and yards up, & sails set; wheels at each side 16ft. diameter.
8. Fresh breeze; square sails in high swell of sea; much rolling and pitching; engines in good order.
9. Good breeze fine weather the fore coal empty in 86 hours; average consumption 6 6lbs avoirdupois per hour.
10. Variable breeze; difficult swell of sea; discovered a second leak in the boiler; steaming, head to wind.
11. Light breezes; less swell; leaks in the boiler increased; afternoon more swell of sea.
12. Fine weather; hailed an English brig, *Jane*, to Demerara from London; rolling strong.
13. Showers; screwed the packings up, and put the same in order.
14. Unsteady breezes; strong rolling; at 11 ½ A.M. stopped the engines, all sails set in the trade-winds.

* EXTRACT from the Log-Book kept on board the Steam Packet, *Curaçao*, Commander J.W. Moll, on her second voyage to Surinam. (Transcribed by Daniel Johnson)

15. Light breezes; repaired the boilers, cleaning the same, and put the engines in order.
16. Fresh breeze; cleaned the engine, and continuing to repair the boiler.
17. Unsteady breezes; repairing the boiler.
18. Light - ditto; ditto - - ditto
19. Light ditto put everything of machinery in order.
20. Light breeze.
21. - ditto - fine weather; long. by observ. W. & E. 48° 38'; Chron. 48° 21'.
22. Good breeze "ditto" more swell of sea; more rolling than before; long. observ. 50° 26'. Chro 50° 32'.
23. Fresh breeze - ditto - - - - - ditto.
24. - Ditto - - ditto - - - - - ditto.
25. - Ditto - thick weather; saw the land at 1 afternoon: at 2½ came to anchor Mot Creek, south.
26. - Ditto - 8½ o'clock in the morning anchored up; steamed the river up; at 2¾ anchored before Parimaribo. The passage has also been made in 25 days, of which 16 hours at anchor. All well on board; no losses whatsoever.

APPENDIX Ja

Appendix, No. 7.—EXTRACT from the Log-Book kept on board the Steam Packet, Curaçao, Commander *J. W. Moll*, on her Second Voyage to *Surinam*.

Dates.	Observed altitude North.	Estimed Longitude of Greenock.	Longitude by Chronometer.	General Course per 24 Hours.	Sailed Distance per 24 Hours.	Greatest Velocity per Hour. Steam.	Sailing.	Steam & Sailing.	Strokes of Pistons per Minute Average.	Stand of Thermometer.	Stand of Barometer.	Hours the Fires lighted.		OBSERVATIONS ON THE WEATHER, AND OTHER OCCURRENCES.
1828: March				4 o'Clock P. M.									W. by S.	- - Fresh breeze, with showers; started at ½ past noon; at 3 outside the Narrows; at 4 o'clock S. E. by E.; 61 men on board, passengers included. Draught of water, 13 ft. 7 inc. and 12 ft. 5 inc.; wheel diameter 15 ft.; on board 360,000 lbs. coals.
1	51° 54′	3° 43′E.	- -	West.	- -	7¼	–	7½	22 & 23	48°	29° 1′	13½	N.N.W. & N.W.	Engines working very smooth.
2	50° 47′	0° 34′E.	- -	W. by S.	136	–	–	8¼	22 & 23	50°	28° 11′	24	N. W. & W.	Strong wind, showers, bad weather; at 4 in the morning St Foreland in sight; at 7½ unshipped the Pilot at Dover.
3	- -	3° 55′W.	3° 55′	W. by S.	182	7	–	7½	23 & 24	53°	29°28′11″	24	N.W. & W.	Light breeze, good weather; at day-break off Start Point; great produce of steam, water good for drinking.
4	48° 0′	7° 10′	7° 37′ 30″	S W. ¼ W.	169	–	–	7½	22 & 23	57°	29° 2′	24	W.N.W. & N.W.	Breeze freshening to tight-reefed top sails; sea very rough; much rolling in the night; sight of the Ouestpant light.
5	45° 35′	10° 6′	10° 13′ 15″	S. W.	172	–	–	8	23 & 24	57°	29° 1½′	24	N.W. & N.	Topgallant breeze; difficult swell of sea; a brig in sight, upon which gained very little.
6	41° 10′	12° 54′	14° 7′	S.W. ½ S.	190	–	–	8	23 & 20	58°	29° 1½′	24	N. & N. E.	Light breeze, fine weather; discovered a small leak in the boiler, S. E.; wheel 17 feet diameter.
7	41° 1′	15° 28′	16° 19′ 30″	S.W. ½ S.	172	–	–	8	20	62°	29° 2′	24	S.E. & S.S.E.	Fresh breeze, fore-top mast and yards up, & sails set; wheels at each side 16 ft. diameter.
8	38° 25′	18° 16′	19° 26′	S.W. ½ S.	207	–	–	9	20 & 21	62°	29° 2′	24	S. S. E. & S.	Fresh breeze; square sails in high swell of sea; much rolling and pitching; engines in good order.
9	36° 10′	21° 6′	22° 18′	S.W. ¼ S	187	–	–	7¾	20 & 21	66°	29°	24	S.S.E. & S.S.W.	Good breeze; fine weather; the fore coal empty in 86 hours; average consumption 66lbs. avoirdupois per hour.
10	34° 5′	23° 40′	24° 45′ 45″	S.W. ½ S.	177	7¼	–	7½	21 & 22	71°	28° 11′	24	S. & S.S.W.	Variable breeze; difficult swell of sea; discovered a second leak in the boiler; steaming, head to wind.
11	32° 4′	25° 41′	26° 40′	S.W. ½ S.	156	–	–	7¾	22	67°	28° 10′	24	W. & S. W.	Light breezes; less swell; leaks in the boiler increased; afternoon more swell of sea.
12	29° 31′	27° 35′	28° 43′	S.S.W. ¾ W	179	7½	–	7	22 & 23	70°	29° 2′	24	S.W. & W.	Fine weather; hailed an English brig, Jane, to Demerara from London; rolling strong.
13	27° 34′	29° 35′	30° 10′ 30″	S.W. ½ S.	133	7½	–	7½	21 & 22	73°	29°	24	N. & N. E.	Showers; screwed the packings up, and put the same in order.
14	25° 25′	31° 38′	32° 20′ 14″	S.W. ½ S	167	–	6	7½	22	71°	29° 2′	11	N. E.	Unsteady breezes; strong rolling; at 11 ½ A. M. stopped the engines, all sails set in the trade-winds.
15	23° 44′	32° 15′	38° 50′	S.W. ½ S	135	–	6¾	–	- -	75°	29° 2′	- -	N. E. by E.	Light breezes; repaired the boilers, cleaning the same, and put the engines in order.
16	21° 59′	35° 8′	36° 15′	S. W.	147	–	9¼	–	- -	74°	29° 1′	- -	E. by E.N.E.	Fresh breeze; cleaned the engine, and continuing to repair the boiler.
17	20° 8′	37° 7′	38° 26′ 30″	S. W.	158	–	8¼	–	- -	75°	29° 2′	- -	E. & E.N.E.	Unsteady breezes; repairing the boiler.
18	18° 17′	39° 6′	40° 48′ 25″	S. W.	158	–	6¼	–	- -	75°	29° 1′	- -	E.N.E. & N.E.	Light - ditto - - ditto.
19	16° 34′	40° 54′	42° 45′ 30″	S. W.	146	–	6¼	–	- -	76°	29° 1′	- -	N.E. ½ E. & N.E.	- ditto - put every thing of machinery in order.
20	14° 57′	42° 23′	44° 25′	S.W. ½ S.	129	–	6½	–	- -	78°	29°	- -	E.N.E. & N.E.	Light breeze.
21	16° 16′	44° 7′	46° 18′	S. W.	142	–	7	–	- -	80°	29°	- -	E.N.E. & N.E.	- ditto - fine weather; long. by observ. W. & E. 48° 38′; Chron. 48° 21′.
22	11° 32′	45° 49′	48° 10′	S. W.	142	–	7	–	- -	80°	29°	- -	E.N.E. & N.E.	Good breeze - ditto - more swell of sea; more rolling than before; long. observ. 50° 26′. Chron. 50° 32′.
23	9° 47′	47° 47′	50° 23′ 30″	S.W. ¼ W.	157	–	7½	–	- -	80°	29°	- -	E.N.E. & N.E.	Fresh breeze - ditto - - - - - - ditto.
24	7° 35′	49° 46′	52° 33′ 30″	S.W. ½ S.	178	–	8½	–	- -	81°	29°	- -	N.E. & E.N.E.	- Ditto - - ditto - - - - - - ditto.
25	6° 16′	51° 7′	54° 47′	S. W.	140	–	–	7½	19 & 20	79°	29°	18	N.E. & E.N.E.	- Ditto - thick weather; saw the land at 1 afternoon; at 2½ came to anchor Mot Creek, south.
26	- -	- -	- -	by the lead	- -	6½	–	–	20 & 21	88°	28° 11′	8	N.E. & E.N.E.	- Ditto - 8½ o'clock in the morning anchored up; steamed the river up; at 2¾ anchored before Parimaribo. The passage has also been made in 25 days, of which 16 hours at anchor. All well on board; no losses whatsoever.
												338½ Hours.		

Average consumption of Coal, 616 to 620 lbs. avoirdupois, per hour.

Parimaribo, 28th March 1828.—The Curaçao Packet, commanded by Lieut. Moll, is a vessel purchased in England by the Dutch government, and employed between Holland and the West Indies; she is 400 tons burthen, and has two 50-horse Engines. She performed her first voyage from Helvetsluys to Surinam in 27 days 11 hours; and from Surinam to Curaçao in 5 days: of this she steamed 11 days 3 hours. She returned from Curaçao to Helvetsluys in 29 days, steaming on the voyage 23½ days. The average consumption of Coal on this voyage was 8½ bushels per hour, or 7·14 lbs. per horse power per hour. We have been favoured with an Extract from the Log-Book of her second voyage out, which was performed in 26 days 8 hours, a copy of which is above. She returned last autumn from the third voyage, commanded by Lieut. Le Jeune, who states this, like the two former voyages, to be most satisfactory, having encountered several severe gales and hurricanes, during which the machinery worked without interruption or accident. The time of her performing this voyage is not stated, nor the consumption of coal. She was then preparing for a fourth voyage.

Extract from the Log Book giving details of the second voyage of the PS *Curaçao. Report from the Select committee on Steam navigation to India, 1834.*

APPENDIX K

Journal of the House of Commons of the Dominion of Canada, 15 March to 23 July 1894, inclusively, Vol. XXVIII, Session 1894, pp. 174-175. Report of the sub-committee on the subject of the *Royal William*

Second Meeting, 1894.

The Joint Committee on the Library of Parliament met a second time in the Chambers of the Speaker of the Senate, on Saturday, May 19th, at 11 A.M.

The Report of the Sub-Committee appointed to consider the erection of a public memorial of the departure of the "*Royal William*" in 1833, was read and adopted.

A copy of the Report, with its Appendices, is annexed hereto by order of the Committee.

The Committee then adjourned.

JOHN J. ROSS,
Speaker.

SPEAKER'S CHAMBERS, May 19th, 1894.

REPORT OF SUB-COMMITTEE ON THE "ROYAL WILLIAM."

The Sub-Committee charged with the duty of inquiring into the proposition laid before them of commemorating, by means of a Memorial Tablet, the departure of the "*Royal William*" from the Port of *Quebec*, in 1833,—the first vessel to cross the Ocean wholly by means of steam,—have the honour to report, as follows:—

That the Sub-Committee have examined the documents prepared and laid before them by Mr. *Sandford Fleming*, C.M.G., whose researches have enabled him to verify the facts contained in the said papers; and that they are of opinion that for the correct statement of a historical fact and to secure for *Canada* the credit attaching to the initiative in so memorable an enterprise as the first crossing of the Ocean wholly by means of steam, some permanent memorial should be placed in some prominent position.

The Sub-Committee are of opinion that a Tablet with a suitable inscription inserted in the wall of the corridor leading to the Library of Parliament, would be a fitting and permanent memorial of an interesting fact in the history of *Canada*; and they recommend that the subject be referred to the consideration of the Finance Minister and the Minister of Public Works, in order that the necessary provision may be made for the Memorial.

Copies of the documents laid before the Committee by Mr. *Fleming* are attached hereto, in order that they also may form part of the public record of so important a fact in the industrial development of modern times.

All of which is respectfully submitted.

JOHN J. ROSS,
Chairman of Sub-Committee.

(*Appendix No.* 1.)

Mr. *Costigan*, a Member of the Queen's Privy Council, laid before the House by command of His Excellency the Governor General,—Annual Report of the Department of Public Printing and Stationery, for the year ended 30th June, 1893, with a partial report for services during six months ending 31st December, 1893. (*Sessional Papers, No.* 16*c*.)

On motion of Mr. *Fraser*, seconded by Mr. *Innes*,

Ordered, That the 51st Rule, and that portion of the 49th Rule of this House, which limits the time for presenting Private Bills, be suspended, in reference to the Bill to incorporate the *Nova Scotia* Steel Company, in accordance with the recommendation contained in the Twelfth Report of the Select Standing Committee on Standing Orders.

Mr. *Haggart*, a Member of the Queen's Privy Council, presented,—Return to an Order of this House, dated 30th March, 1894, for a Return showing the sums of money paid to Mr. *A. F. Wood*, Government valuator, for services, maintenance and transport during the years 1891-1892-1893, and the particular pieces of land or other property valued during those years. (*Sessional Papers, No.* 73.)

Sir *Charles Hibbert Tupper*, a Member of the Queen's Privy Council, presented,—Return to an Address to His Excellency, dated 30th March, 1894, for a Return of all Papers, Correspondence, Telegrams, Reports to and Orders in Council, together with all Departmental Orders relating to the discharge from office as Superintendent of *St. Paul's Island*, and keeper of *Ingonish Island*, of Mr. *Samuel Campbell*, or relating to his superannuation allowance. (*Sessional Papers, No.* 33*d*.)

Mr. *Ives*, a Member of the Queen's Privy Council, presented,—Return to an Address to His Excellency, dated 16th April, 1894, for copies of all Orders in Council in force in *Canada* (Provinces of Lower *Canada* and Upper *Canada*) in 1858, concerning any drawback or bounty with respect to the building of Canadian ships, barques and other vessels; and also, all Orders in Council amending the same, or concerning the same from 1858 up to the beginning of the Confederation. (*Sessional Papers, No.* 74.)

Mr. *Mills* (*Annapolis*), moved, seconded by Mr. *Sproule*, and the Question being put, That the Order of the House for the second reading of the Bill from the Senate, intituled: "An Act for the relief of *Joseph Thompson*," be discharged, and the said Bill referred to the Select Standing Committee on Standing Orders; the House divided: and it was resolved in the Affirmative.

Ordered, That Mr. *Ives* have leave to bring in a Bill to amend and consolidate the Acts respecting the *North-west* Mounted Police Force.

He accordingly presented the said Bill to the House, and the same was received and read the first time; and ordered to be read a second time on Monday next.

Glossary

Aviso	Dispatch boat.
Blockade	To block entry or exit from a place by surrounding it with troops or ships.
Blow off	Use a safety valve to blow off steam.
Bring in	Bring in a vessel to a wharf berth.
Capstan	A machine for moving or raising heavy weights consisting of a vertical drum which can be rotated, usually manually, and around which cable or rope is turned.
Carronade	A gun with a short barrel and wide mouth.
Chadron	A measure of coals consisting of 36 bushels.
Charter	To hire a ship by contract.
Clinkers	Incombustible matter formed in a furnace from impurities in the coal.
Combings	Raised edge around ship's hatches to keep water out.
Consign	To deliver to another by a formal agreement.
Escape valve	A valve that releases air, steam, etc., above a certain pressure.
Fashion pieces	A timber that frames the shape of the stern.
Flank	The side of an army, or of a division of an army.
Floating dock	Trough-shaped structure used to lift ships out of the water for inspection or repairs.
Forecastle deck	Short raised deck at fore end of ship; fore of ship under main deck.
Fore topmast	The mast erected above the lower mast on the forward mast of a vessel.
Grate	An iron frame used to hold and burn coal.
Gunwale	The upper edge of a ship's side.
Iberian Peninsula	The peninsula comprising Spain and Portugal.
In frame	A vessel is in frame when it is completed with the exception of the ceiling and the outside planking.
Ironclad	A warship covered wholly or partially with thick iron plates.
Jolly boat	A ship's boat that is smaller than a cutter, typically housed at the stern of the ship.

Keelson	An internal longitudinal timber or line of timbers, mounted on top of the frames along the centerline of the keel, that provided additional longitudinal strength to the bottom of the hull.
Knee	A piece of timber with a knee or angle in it.
To Lay to	To bring a ship to a stop in open water by setting sails in apposition to each other.
Man-of-war	A warship or frigate.
Mercenary	A soldier who is hired into Foreign Service.
Millwright	A person who designs or builds mills or who maintains mill machinery.
Mould loft floor	A mould loft's large covered wooden floor area used for laying down plans of a ship.
Port	The left-hand side of a ship when facing forward.
Quarter badges	An ornamental window-surround, or a representation of a window, at the stern of a vessel.
Raft-ship	A ship built of square timbers.
Re-sheath	To re-cover the hull with metal plates.
Roundhouse	A cabin on the after part of the quarter-deck of a ship.
Scroll head	Ornamental molding used in place of a figurehead.
Shipbroker	A broker who obtains insurance or performs other business for ships.
Side beam engines	Developed for use on ships, as they took up less room than regular steam engines.
Skiff	A light rowing boat or sculling boat, typically for one person.
Spar	A thick, strong pole such as is used for a mast or yard on a ship.
Specie	Gold or silver coin in contradistinction to paper money.
Squadron	A detachment of ships of war employed on a particular mission under a flag officer.
Starboard	The right-hand side of a ship when facing forward.
Sterling	British coinage.
Surface condenser	A heat-transfer device used to condense a vapor.
Transom	A horizontal beam reinforcing the stern of a boat.
Transport	A large ship used to carry troops or stores.
Treenail	Long wooden pin used to fix ship's planking to its timbers.

Bibliography

MANUSCRIPT SOURCES

Bibliothèque et Archives nationales du Québec, Centre d'archives de Montréal (ANQM)

Greffe of the notary H. Griffin (CN601).

Contract between Charles Fry and William Stevenson on behalf of the Quebec and Halifax Steam Navigation Company, 10 March, 1831.

Contract between Bennet and Henderson and the Quebec and Halifax Steam Navigation Society, 19 March, 1831.

Bibliothèque et Archives nationales du Québec, Centre d'archives de Québec (ANQQ)

Black, George. "Letter Book." AP-P-203, ANQQ.

Greffe of the notary Archibald Campbell (CN301)

Joseph Wilson Henry, Affidavit, 26 February, 1891.

Greffe of the notary F.-X. Garneau, 25 January, 1831.

Greffe of the notary Lauchlan Thomas McPherson (CN301).

Contract Bennet and Henderson with the Quebec and Halifax Steam Navigation Company, 28 August, 1830.

Contract George Black with the Quebec and Halifax Steam Navigation Company, 3 September, 1830.

Contract for building the cabins of the steamship between Joseph Andrew with the Quebec and Halifax Steam Navigation Company, 22 January, 1831.

Ratification of contract made in Montreal on the 19 March 1831, between Charles Fry and William Stevenson, 23 April, 1831.

Procuration Quebec and Halifax Steam Navigation Company to William Mortimer of Pictou, April 17, 1833.

Literary and Historical Society of Quebec.

Quebec shipping Registers 1831-1833, LAC.

Register of steamship *Royal William,* 22 August, 1831.

Register of steamship *Royal William,* 18 May, 1833.

Chicago Historical Society

Letter to Society from Archibald Bennet, 31 October, 1904.

Letter to Society from Geo. G. Goudie, 20 November, 1904.

Letters from James Goudie to L. Hagadorn, President of the Society, 22 June and 4 October, 1888.

Royal Museums Greenwich – National Maritime Museum, Caird Library, Royal Mail Steam Packet Company.

Toronto Public Library, Special Collections

Kivas Tully, correspondence and Scrapbook.

Vancouver Maritime Museum Archives

Specification of the proposed steam vessel. July 1, 1830.

Other correspondence

John Lowe to his parents, 9 October, 1831. Fred Colby papers, Grosse Pointe Farm, Michigan, neg. C031338, LAC.

James Berry to his uncle, 15 November, 1831. Berry family papers, Newport-on-Tay, Scotland.

Samuel Cunard to Rupert D. George, May 7, 1833, file 3018-20, Phyllis Blakeley Collection, Nova Scotia Archives.

John MacDougall to William King, 16 November, 1833; to Robert Christie, 10 August, 1853. (Both in *Report of the Secretary of State, 1894).*

James Goudie to James W. Stewart, 22 December, 1877 in *Morning Chronicle,* of unknown date; James Goudie to Archibald Campbell, 17 February, 1891, *Report of the Secretary of State, 1894.*

Alfred Jephson to Quebec Literary and Historical Society, 30 March, 1891. Literary and Historical Society of Quebec, ANQQ.

The author with Elizabeth Wickstead, 2014.

PRINTED SOURCES

Government

Journals of the House of Assembly of Lower Canada. 24 February, 7 and 22 March, 1825; 4, 6, 26 and 30 March, 3 September, 1830; 28 January, 31 March, 29 November, 1831; February, 1832; 22 March, 1849.

Journal & Proceedings of the House of Assembly, Halifax, 2 and 5 April, 1825, 4 and 30 March, 1830; 12 and 13 February, 1835.

Report from the Select Committee on Steam Navigation to India: With the Minutes of Evidence, Appendix and Index, Appendix 7, Extract from the Log Book kept on board the steam packet Curaçao. House of Commons, 1834.

Report of the Secretary of State of Canada for the Year ended 31st December, 1894.

Royal Gazette 13 September, 1831.

Newspapers & Magazine (Canadian)

Acadian Reporter June, 1830, 3 September, 1831.

Montreal Gazette 21 August, 1822; 3 September and 22 November, 1830; 6 and 16 April, 3 May, 14, 19, 21 and 30 July; 2 and 18 August,1831; 3 and 6 August, and 16 March, 1833.

Pictou Colonial Patriot, 13 and 20 August, 1833.

Quebec Gazette 20 October, 1825; 18 February, 1830; 25 or/and 27 and 28 April, 19 May; 14, 18, 21, 30 July, 3, 6 and 18 August, 1831; 10 and 17 July, 14 (24) September, 1833.

Quebec Mercury 23 and 30 April, 1831; 16 July, 1833.

Quebec Morning Chronicle November, 1833; Letter to the editor from James W. Stewart, December, 1877.

Royal Gazette, 13 September, 1831.*Sydney Herald.* New South Wales, 29 November, 1832.

The Canadian Magazine and Literary Repository, Vol. 4, 1825.

The Daily Witness, 20 March, 1897.

Newspapers and Journals (other)

Aberdeen Journal, December, 1831.

American Popular Science Monthly, January, 1893.

Berkshire Chronicle, 14 September, 1833.

Kentish Gazette, Deal, 17 September, 1833.

Leeds Intelligencer, 4 August, 1825.

London Gazette, 10 September, 1833.

Morning Post, London, 11 September, 1833

Nautical Magazine and Naval Chronicle 10 July, 1845 and others

Times (London), 12 August, 1833; 11 September, 1833.

Toronto Daily Globe,

Washington National Intelligencer, 4 November, 1833.

Western Flying Post, Sherborne and Yeovil, 16 September, 1833.

Books and Articles

BÉLAND, Mario. "Le Royal William, un vapeur légendaire de Québec." *Cap-aux-diamants,* No.74, été, 2003.

—"Le *Royal William* (1831) 1ère partie." *Neptunia* 231, septembre, 2003.

— "Le *Royal William* (1831) 2e partie (fin)." *Neptunia* 232, décembre, 2003.

BOILEAU, John. *Samuel Cunard: Nova Scotia Master of the North Atlantic.* Halifax: Formac Publishing, 2006.

BONSOR, Noel, Reginald PIXELL. *North Atlantic Seaway; An Illustrated History of the Passenger Services Linking the Old World with the New_* (2nd ed. Jersey, Channel Islands: Brookside Publications, 1975).

BRADLEE, Francis B. C. "The First Steamer to Cross the Atlantic. The record of the Steamship Curaçao of the Royal Netherlands Navy." Salem, Mass: The Essex Institute, 1922.

BROWN, David K. *Before the Ironclad: Development of Ship Design, Propulsion, and Armament in the Royal Navy, 1815–1860.* London: Conway Maritime Press, 1990.

BUCHWALD, Jed Z. FOX, Robert, eds. *The Oxford Handbook of the History of Physics.* Oxford: Oxford University Press, 2013.

BUGGEY, SUSAN and DAVIES, Gwendolyn, "McCulloch, Thomas." *Dictionary of Canadian Biography,* vol.7. University of Toronto/ Université Laval, 1988, revised 2015.

CALABUIG, D. Jos. Lledo. *Buques de Vapor de la Armada Española del vapor de ruedas a la fragata acorazada, 1834-1885. Madrid:* Aqualarga, 1998.

CAMPBELL, Archibald. "The Royal William, Pioneer of Ocean Steam Navigation." *Transactions of the Literary and Historical Society of Quebec* (New Series). No.20, Sessions, 1889-1891.

CHRISTIE, Robert. A history of the late province of Lower Canada., Parliamentary and Political, Vol.5. Quebec, 1854.

DELGADO Bañón. LUIS M. *El vapor a Ruedas Isabel II.* Barcelona:Editoroial Noray, 2013.

DENNISON, Merrill. *The Barley and the Stream.* Toronto: McClelland & Stewart Limited, 1955.

DUBUC, Alfred. "Molson, John, sr." Alfred Dubuc, *Dictionary of Canadian Biography,* vol. VII.

FAULKNER, Christopher. "Some Particulars Concerning the Lauzon Ferry and its Token." *Canadian Numismatic Research Society (New Series) Transactions,* vol.2, 2014.

FLEMING, Sandford. *Steamboat Service between Canada and Great Britain.* Kingston: Offprint, Queen's Quarterly, 1894.

FRY, Henry. *The History of North Atlantic Steam Navigation with some account of Early Ships and Shipowners.* London: Sampson Low, Marston & Co., 1896.

GUILLEN, Julio F. "El bloqueo del Cantabrico." *Boletín de la Real Academia de la Historia,* vol. 124, #2, April/June. 1940 pp. 392-414, Madrid: Fortanet, 1951.

JEPHSON, Harriet J. *Notes of a Nomad.* London: Hutchinson, 1918.

KENNEDY, John. *History of Steam Navigation.* Liverpool: Charles Birchall, 1903.

KER, Robert. "The Pioneer of Atlantic Steamships." *The Canadian Magazine of Politics, Science, Art and Literature,* Vol. XXIX, May to October, 1907. Toronto: The Ontario Publishing Company Limited, 1907.

LANGLEY, John G. *Steam Lion – A biography of Samuel Cunard.* Halifax: Nimbus Publishing, 2006.

LEWIS, Walter. "The First Generation of Marine Engines in Central Canadian Steamers, 1809-1837," *The Northern Mariner,* VII, no. 2 April, 1997.

MACKEY, Frank. Steamboat Connections: Montreal to Upper Canada, 1816-1843. Montreal: McGill-Queen's University Press, 2000.

MARCIL, Eileen Reid. *The Charley-Man: A History of Wooden Shipbuilding at Quebec, 17631893.* Kingston: Quarry Press, 1995.

— *On chantait "Charley-Man." La construction de grands voiliers à Québec de 1763 à 1893.* Quebec: GID, 2000.

L'extraordinaire exploit de Charles Wood : Le Columbus *et le* Baron of Renfrew. Quebec: Les Éditions GID, 2011.

McCAIN, Charles W. *History of the* SS *Beaver, Being a Graphic and Vivid Sketch of This Noted Pioneer Steamer.* 1894. Reprint. London: Forgotten Books, 2013. pp. 90-1.

MOLSON, Karen. *The Molsons: Their Lives and Times: 1780-2010.* Willowdale, Ontario: Firefly Books, 2001.

OSBORNE, Eric W. *Cruisers and Battle Cruisers: An Illustrated History of their Impact.* Oxford: ABC- CLIO, 2004.

PARDO SAN GIL, Juan. "Las operaciones navales en las Guerras Carlistas." *ISAS Memoria. Revista de Estudios Maritimos des Pais*

Vasco, 5, pp. 433-466. Donostia-San Sebastian; Untzi, Museoa-Naval, 2006.

PATTERSON, George. A History of the County of Pictou, Nova Scotia. p. 395.

POOLE, Henry, HONEYMAN, David. *Notes on the Coal Fields of Pictou.* Delhi: Facsimile Publisher, 2015. First published 1860.

ROWLAND, K. T. *Steam at Sea: A History of Steam Navigation.* New York, N.Y.: Praegar Publishers Inc., 1970 and K.T. Rowland, London, England.

ROYAL NAVAL EXHIBITION. *Official Catalogue and Guide.* Opened at Chelsea May 2, 1891, London: W.P. Griffith & Sons Limited.

SMITH, Captain Edgar C. *A Short History of Naval and Marine Engineering.* Cambridge: University Press for Babcock & Wilcox Limited, 1937.

SMITH, Crosbie. *Coal, Steam and Ships.* Cambridge: Cambridge University Press, 2018.

SPRATT, Philip. *Transatlantic Paddle Steamers.* Glasgow: Brown, Son & Ferguson, 1951.

STAMMERS, M.K., *Steamboats.* Aylesbury, Bucks: Shire Publications, 1986.

TALBOT, Edward Allen (ed.). *Five Years Residence in the Canadas.* London: Longman, Hurst, Rees, Orme, Brown & Green, 1824.

VAN NOUHUYS, J, W. *De Eerste Nederlandsche Transatlantische Stoomvart in 1827 Van Zr Ms Stoompakket* Curaçao, Eerste deel het journaal. Gravenhage: Martinus Nijhoff 1927 and 1951.

WALKER, James. *An investigation of the claims put forward on behalf of the Canadian steamer* "Royal William." With notes by Sir C. Purcell Taylor. London: Swan Sonnenschein & Co., ILM. Paternoster Square, 1898.

WILCKEN, Patrick. *Empire Adrift*: *The Portuguese Court in Rio de Janeiro 1808-1821.* Bloomsbury Publishing PLC, 2005.

WILLIAMS, David M. and ARMSTRONG, John. "Promotion, speculation and their Outcome: The Steamship Mania of 1824-1825." *Aslib Proceedings New Information Perspectives,* 60:6, 2008, pp. 642.

—*The Perception and Understanding of New Technology: A Failed Attempt to Establish Transatlantic Steamship* Liner Services 1824-1828." *The Northern Mariner/le marin du nord, XVII No. 4,* October, 2007, pp.41-56.

WILSON, George H. "The application of steam to St. Lawrence valley navigation 1809-1840". McGill Digital Library, 1961.

WOOD, William. "The record-making "*Royal William." Canadian Geographic Journal,* VII (2), Aug. 1933, 153-63.

WURTELE, F.C. "SS *Royal William,* the pioneer of transatlantic steam navigation." App. G, *Report of Secretary of State of Canada for 1894*, Ottawa, 1895. p. 60.

Index

(For names of the 235 founding shareholders, please see list in Appendix F, p. 108)